Joshua Rose

The Slide Valve Practically Explained

Joshua Rose

The Slide Valve Practically Explained

ISBN/EAN: 9783337379544

Printed in Europe, USA, Canada, Australia, Japan

Cover: Foto ©berggeist007 / pixelio.de

More available books at **www.hansebooks.com**

THE

SLIDE VALVE

PRACTICALLY EXPLAINED.

EMBRACING

SIMPLE AND COMPLETE PRACTICAL DEMONSTRATIONS OF THE OPER-
ATION OF EACH ELEMENT IN A SLIDE-VALVE MOVEMENT, AND
ILLUSTRATING THE EFFECTS OF VARIATIONS IN THEIR
PROPORTIONS, BY EXAMPLES CAREFULLY
SELECTED FROM THE MOST RECENT
AND SUCCESSFUL PRACTICE.

BY

JOSHUA ROSE, M.E.,

"The Complete Practical Machinist," "The Pattern Maker's
Assistant," etc. etc.

ILLUSTRATED BY 35 ENGRAVINGS.

PHILADELPHIA:
HENRY CAREY BAIRD & CO.,
Industrial Publishers, Booksellers, and Importers,
810 WALNUT STREET.
1880.

PREFACE.

The object of this book is to present to practical men a clear explanation of the operations of a slide valve under the conditions in which it is found in actual practice. The author believes that its clear and concise treatment of the subject will leave nothing to be desired, even by those who begin it in entire ignorance of the subject matter.

CONTENTS.

	PAGE
Simplest form of valve	5
Lead and its effects	7
Diagram of simple valve action	11
Steam *lap* explained	13
Diagram showing effects of steam lap	17
The angularity of the connecting rod and its effect on the valve action	18
The eccentric piston and crank movements	21
Exhaust *lap* explained	25
Diagram showing effects of exhaust lap	28
Clearance explained	30
Effects of clearance	31
Valve travel considered	32
Effect of over travel on the position of the eccentric and port openings	34
Point of cut-off	37
The rock shaft and its effect on the position of the eccentric and on the distribution of the steam	40
Examples from practice	44
Comparison of the merits of the various examples	57
Comparison of port areas	58
" compression	59
" expansion	60
" exhaust areas	61

CONTENTS.

	PAGE
Comparison of valves of exhaust areas	62
" ports to their duty	64
" areas of wearing surface	67
Rules for areas of steam ports	69
Shapes of ports	72
Power required to operate valve	76
Conditions as affecting friction	77
Warping of valves	80
Spring of valves from pressure	81
Coefficient of friction	84
Practical experiment on friction	87
Size of valve as affecting friction	88
Lubrication of slide valves	88
Ascertaining with a pair of compasses the result to be obtained from a given slide valve	89
To ascertain point of full steam port opening	90
" position of piston when port is full open	90
" point at which valve begins to close	90
" point of cut-off	92
" amount of expansion	92
" point of release	92
" point at which exhaust begins	93
" point of full exhaust opening	93
" point where cushioning began	93
" amount of cushioning	93
To set a single eccentric without moving the engine	95
" a double eccentric without moving the engine	96
" eccentrics when the rod is not in line with the cylinder bore	97

THE SLIDE VALVE.

WITHOUT LAP.

The simple term *slide valve* is understood to mean a single valve such as is shown in Fig. 1. Devices which have a second valve for the purpose of cutting off the steam supply or for the exhaust, are distinguished by a second appellation denoting the object of the design or the action of the valves, as "cut off slide valve," "automatic cut-off slide valve," etc. etc. The slide valve is in universal use for locomotives and small stationary and boat engines, but has been largely replaced in large stationary engines by automatic cut-off valves.

In some cases a steam engine cylinder has its slide valve motion constructed in two divisions, a steam port and an exhaust port at each end of the cylinder, the two valves being connected together by a rod. The object of this arrangement is to obtain short steam passages; but since the action of the valve is precisely the same as if the two valves were consolidated into one, as in all our examples, no attention need be given to them.

The simplest form in which the slide valve is made, is that shown in Fig. 1, in which A represents the steam port for one, and B, the steam port for the other end of the cylinder, while

C represents the cylinder exhaust port; DD, the bridges; E, the slide valve, and F, the valve exhaust port. The valve is shown in the position in which it stands when the piston is at

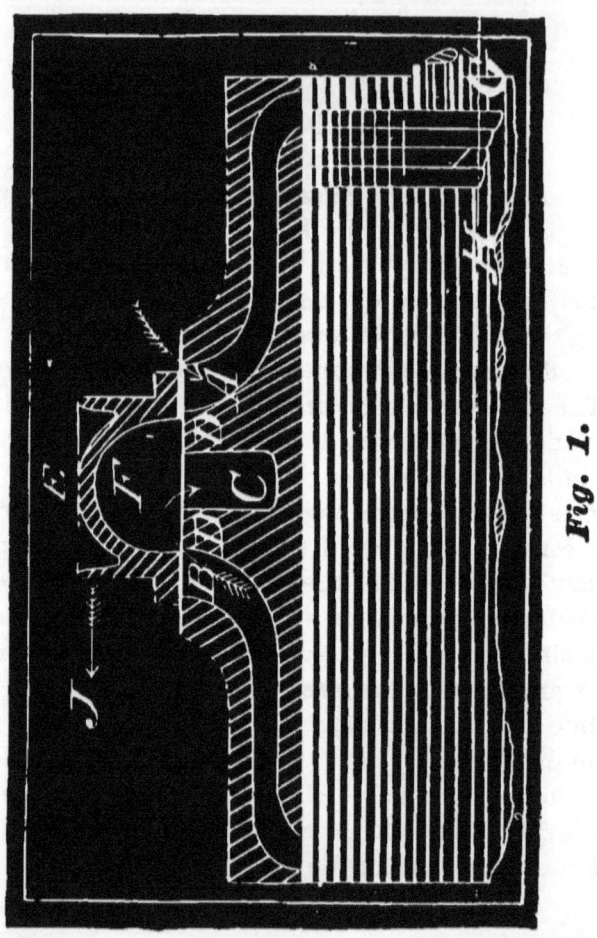

Fig. 1.

the end, G, of the cylinder; the port A, acting as a steam port that is to admit steam from the steam chest into the cylinder. The steam in the cylinder in the side, H, of the piston (which steam propelled the engine the previous stroke) now finds egress through the port, *B*, and thence through *F* and *C* to the exhaust pipe.

The valve after traveling in the direction of the arrow, *J*, until it leaves the port, *A*, full open, reverses its motion and travels back, closing the port, *A*, and opening, *B*, to the steam in the steam chest. It will be noted that there is shown a slight opening left by the valve to the port, *A*, and the amount of this opening when the piston is at the exact end of its stroke is called the *lead* of the valve. The action of *lead* is threefold; first, it admits steam to the piston before it has arrived at the end of the stroke, and this steam acts as a cushion, causing the piston, etc., to reverse its motion easily. Secondly, it assists the admission of the steam, tending to permit the steam passage to become supplied with steam at the steam chest pressure by the time the piston reverses its motion; and, thirdly, it assists the exhaust in its early part, which is of especial importance to a valve such as is shown in Fig. 1.

If the valve had no lead it would stand so as to just close the two ports, *A* and *B*, when the piston stood at either end of the stroke, and then while the piston made a stroke the valve would move so as to first open and then close the requisite port. Suppose, for example, the piston shown in Fig. 1 were to make a stroke; while it was doing so, the valve would move in the direction of the arrow, *J*, until it left the port, *A*, wide open, whereupon, it would travel back and reclose that port, arriving at the precise position from which it started at the same instant that the piston terminated its stroke. It will be noted that precisely as the port, *A*, opens as a steam port, the port, *B*, opens as an exhaust port, and *vice versa*, so that while *C* and *F* always act as exhaust ports, *A* and *B* act alternately and respectively as steam and exhaust ports, the lead of the valve acting the same at each end of the stroke, that is, opening the steam port on one side and the exhaust port on the other side of the piston in advance of the piston arriving at the terminal of its stroke. The lead of the valve is determined by the position of the eccentric. In Fig. 2, $a\,A$ represents the center line of the crank, and c the throw line of the eccentric, the line; E, representing an imaginary line, standing at a right angle to $a\,A$. Now if the eccentric stood so that its throw line was at a right angle to the center line, $a\,A$ of the crank, it

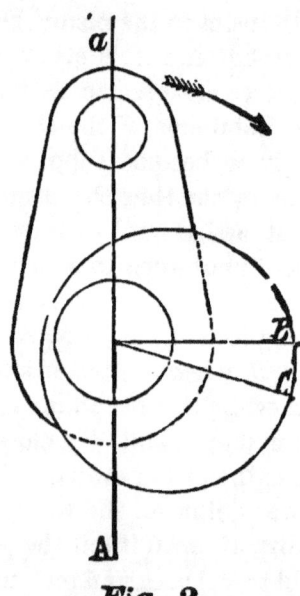

Fig. 2.

would have no lead, or angular advance, as it is sometimes termed, and hence, the valve would have no lead. Supposing, however, the crank in Fig. 2 to revolve in the direction of the arrow, by moving the eccentric forward (that is in the direction in which the crank is to revolve) so that its throw line stands in advance of a line intersecting the center of the crank-shaft, and at a right angle to the center-line, $a\,A$, of the crank a valve may be given any desired amount of lead.

The angular advance or lead of the eccentric shown in Fig. 2 being the distance between c and E. In American locomotive practice, the amount of lead given varies from about 3-32d to 3-16th inch, while in English practice it is generally more, ranging from 3-32d to $\frac{1}{4}$th-inch. In stationary cut-off engine practice, from 1-64th to $\frac{1}{8}$th-inch lead is the usual proportion, and in many cases not more than 1-64th-inch is employed. If a valve is given exhaust lap, a portion of the exhaust steam is enclosed in the cylinder, and acts both to cushion the piston and fill the passages with steam at a pressure, the difference between its action and that of the *lead* being that the former renders the exhaust less free, while the latter assists it. Since, however, a valve having no steam lap is apt (unless its movement is a very slow one) to have a cramped exhaust, lap on the exhaust side is inadmissible in such valves and the subject need not, in this connection, be discussed.

The objections to the use of a valve such as shown in Fig. 1, are that it does not cut off the steam supply before the piston has arrived at the end of its stroke, and the steam in the cylinder is not used expansively, and, therefore, not economically. Then, again, the admission and exhaust of the steam to and from the cylinder, takes place slowly, with the result that neither occurs freely until the piston has traveled some distance. To demonstrate the effect of such a valve it will be well to introduce the form of diagram it is intended to employ to illustrate the operation of the various forms of valves to be hereafter considered.

Let us suppose, then, that we have in our possession an engine of 24-inches stroke, the connecting rod being three times the length of the stroke; the valve elements being as follows:—Width of steam ports, 1-inch; width of bridge $\frac{3}{4}$-inch; width of exhaust port 1-inch; lap 0, lead 1-16th inch; travel of valve 2 inches. Then starting from the end of the stroke nearest to the crank, we move the engine piston an inch, and measure how much the steam-port is open ; then after moving the piston another inch of the stroke, we again measure the amount the steam-port is open, making in each case a note of the data thus obtained, and continuing the operation through a full revolution of the engine, and noting down the position of the valve on both the steam and exhaust ports respectively. From the data thus obtained, we construct our diagram, as follows :—We draw the line, A, B, in Fig. 3, which represents the length of the engine stroke ; this line we divide off into as many equal divisions as there are inches in the piston stroke, so that each line denoting a division will denote an inch of piston movement. Above this line, AB, we draw the parallel line, $c\ c$, the distance between the two lines being the exact width of the steam ports of our engine. The line, $D\ D$, is next drawn, parallel also with $A\ B$, but below it to an amount equal to the width of the steam port. The lines at right angles to the line, AB, and representing the inches in the engine stroke, being drawn to intersect, respectively, the three parallel lines $A\ B$, $c\ c$, and $D\ D$, we take our data of the piston and valve movements, and proceed as follows :—Beginning at A, we make a mark above it, a distance equal to the amount of *lead* on the valve of our engine, then passing to the next line (above the line $A\ B$,) we make thereon another mark, its distance or height from $A\ B$, being the amount that our data show the valve was open when the engine piston had moved an inch. This process we continue until we arrive at the 24th line, and then a line drawn through all the marks so made will show at a glance how the port was opened for the admission of the steam. Then turning to our data, we mark, in like manner, on the lines from 1 to 24, respectively, but below the line $A\ B$, and beginning at the right hand, the width the exhaust port stood open from the 1st to the 24th inch of the exhaust, and drawing a line through these points will denote exactly how the port opened and closed during the exhaust of the steam. Thus, then, we

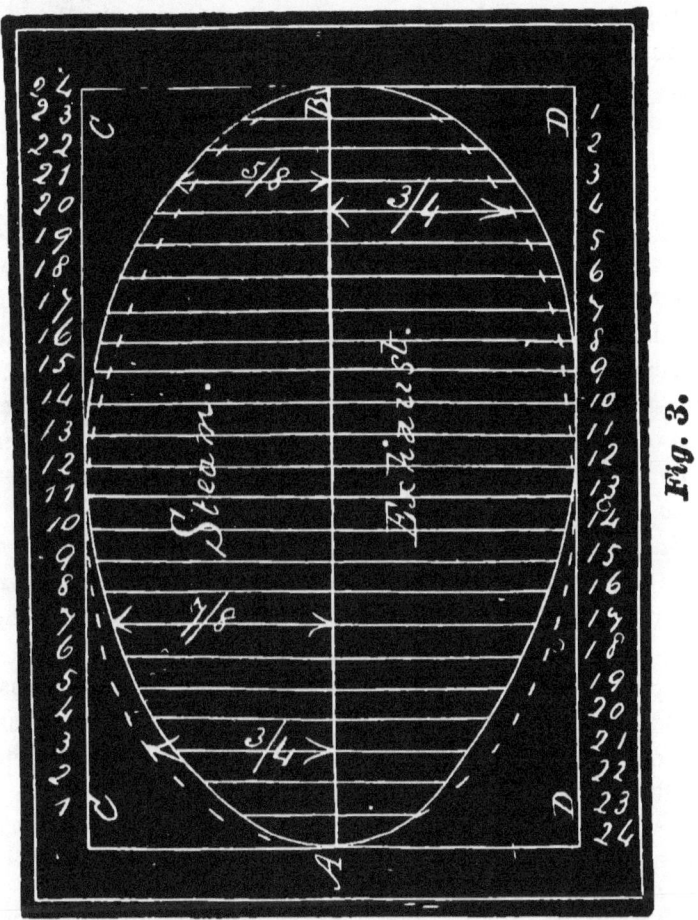

Fig. 3.

have a record of the admission and exhaust for one stroke of the engine, or, in other words, of the admission and exhaust of the steam through the port nearest to the crank end of the cylinder. We then go through the same process, using the data obtained by the other stroke of the engine piston; but in this case we leave the dots intact instead of drawing a line through them, and thus we not only distinguish one diagram from the other, but we are afforded an excellent opportunity to perceive and compare the difference in the action of the valve during one stroke as compared with the next. Thus on referring to Fig. 3, we read that during one stroke, the port did not open full as a steam port until the piston had traveled more than eleven inches of the stroke, while during the other stroke it stood full open at the tenth inch of the stroke. Then we find that the exhaust was full open for one stroke at the ninth inch of the piston movement, and for the other stroke at the thirteenth inch. We may also readily observe that on the stroke in which the valve opened the slowest for the admission, it opened the quickest for the exhaust of the steam, and if we desire to know how wide open the port stood at any particular inch of the stroke, we have but to measure the length of the vertical line referring to that inch of the stroke. For example, when the piston was moving away from the crank, the port at the seventh and twenty-first inches of the piston movement stood open ⅞-inch and ⅝-inch, respectively (as marked upon the diagram), or during the other stroke of the piston the port stood open as a steam port ¾-inch, when the piston had moved 3 inches, and stood open ¾-inch as an exhaust port, when the piston had moved 4 inches of the return or exhaust stroke.

STEAM LAP, OR LAP ON THE STEAM SIDE OF THE VALVE.

In all engines in which the slide valve is operated by a revolving eccentric, the valve should have what is termed steam lap, for although without this lap the steam at about steam-chest pressure will follow the piston for the full length of the stroke, as shown in Fig. 3, yet the engine will exert less power than it would if by means of steam lap the steam supply were cut off at some point before the piston arrived at the end of the stroke. At what particular point this will be in any given engine will depend upon the speed at which the engine runs, but in any event the faster the engine runs the earlier in the stroke should the steam supply be cut off to obtain the maximum of power from the engine. Now especially is this the case if the valve have but little or no lead. The reason of this is, that with a minimum of lead and no steam lap, the steam on the exhaust side of the piston cannot escape quickly enough to prevent its exerting a back pressure on the piston.

The *lap* of a valve is the amount by which it overlaps the cylinder steam ports when it stands in mid-position; thus, in Fig. 4 the lap is the distance from A to B on one side and from C to D on the other side.

When the word *lap* is used unaccompanied by the terms steam or exhaust, it is understood to refer to lap on the steam side of the valve (as shown in Fig. 4), because while nearly all valves have steam lap, a majority of them have no exhaust lap. So, likewise, the edges E, E, of the valve (Fig. 4) are called the steam edges, because they operate to admit or cut off the steam to and from the cylinder ports, and are at all times isolated from the exhaust ports. The edges F, F (Fig. 4) of the valve are called the exhaust edges because they operate to open or close the exhaust of the steam from the cylinder. The *lap* is usually measured on one side of the valve only; thus, if (in Fig. 4) the valve measured an inch from A to B and an inch from C to D, it would be said to have an inch of lap, or an inch of steam lap, both terms signifying the same thing.

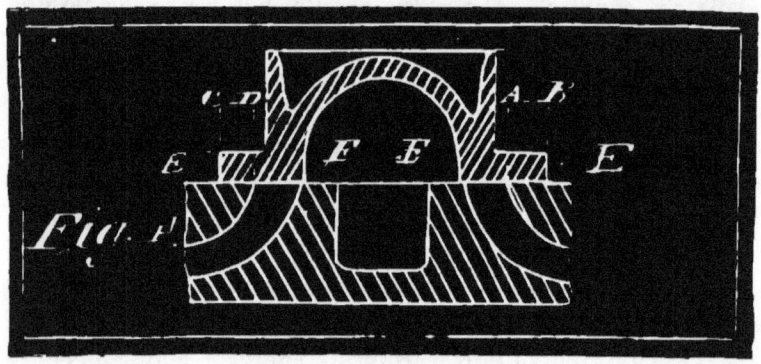

The object of the steam lap is to cut off the supply of steam from the steam chest into the cylinder before the piston has completed its stroke, and the point in the stroke at which the supply is cut off, is called the point of cut-off.

By using the steam expansively a more useful effect is obtained from it than would otherwise be the case. Suppose for example that the steam is exhausted from a high pressure engine cylinder at a pressure of 40 lbs. per square inch, and that the pressure of steam in the boiler was 75 lbs. per square inch, then it is evident that the fuel consumed in order to convert the feed-water into steam up to a pressure of 40 lbs. per square inch (the exhaust pressure) is not represented by any power imparted to the piston. Another way to view the gain due to expansion is to suppose a piston having a stroke of 24 inches to receive steam from the steam chest during the first 12 inches of its stroke only, then while it is traveling the last 12 inches the steam in the cylinder will expand and still impart power to the piston, and this power constitutes the gain due to working the steam expansively, since it was obtained without the expenditure of any steam from the steam chest other than that necessary to fill the the cylinder during the first 12 inches of the piston stroke. It follows, then, that the steam is used the most economically when it is exhausted from the cylinder at a pressure equal to that of the atmosphere.

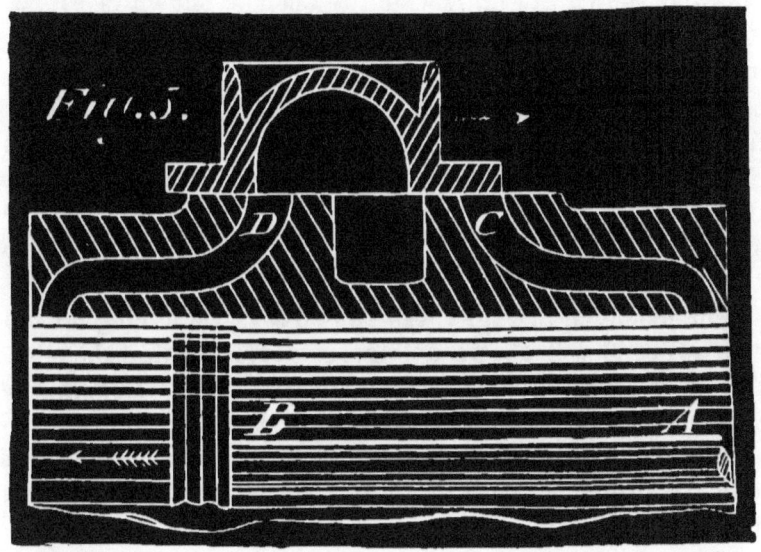

In Fig. 5, the valve is shown in the position in which it cuts off the steam supply to the cylinder, the piston and the valve traveling respectively in the direction denoted by the arrows, and the steam occupying that portion of the cylinder from A to B.

Another and important object gained by giving a valve

steam lap is the obtainance of a free exhaust, which is of great importance, especially in a fast running engine. Whatever the amount of steam lap is, to that extent will it cause the valve (supposing it to have no exhaust lap) to leave the exhaust port open when the piston is at the end of its stroke. If, for instance, we suppose the piston in Fig. 5 to be at the end, A, of the cylinder, then the valve would be in the same position, but traveling in the opposite direction, to that denoted by the arrow, and the port C, would be on the point of opening as a steam port, while the port, D, would be acting as an exhaust port, and the latter, instead of being, closed as was the valve shown in Fig. 1, would be open, as shown in Fig. 5.

In addition to this, however, the valve having lap must, in order to let the steam ports open full, necessarily have a proportionate increase of travel, and this again helps to give freedom to the exhaust steam by keeping the exhaust port open during a greater portion of the stroke. In like manner the increase of valve travel causes the valve to open the steam port quicker, giving a more ready supply of steam to the cylinder.

For example, in Fig. 1 was shown a valve having no lap and in Fig. 3 was shown a diagram demonstrating the opening or closure of the steam and exhaust ports. In Fig. 5 is shown a similar valve with steam lap added. The proportions being: Width of steam ports 1 in.; width of bridges $\frac{3}{4}$ in.; width of exhaust port 2 in.; steam lap $\frac{3}{4}$ in.; exhaust lap 0; travel of valve $3\frac{1}{2}$ in.

The only variations between these dimensions and those of the valve shown in Fig. 1, are that $\frac{3}{4}$-inch steam lap has been added, and, as a consequence, the cylinder exhaust port has been widened.

In Fig. 6 is shown a diagram of the above valve, obtained, as before, by measuring the ports at each inch of piston movement, the full lines being for the cylinder port farthest from the crank, and the dotted lines being for the port nearest to the crank.

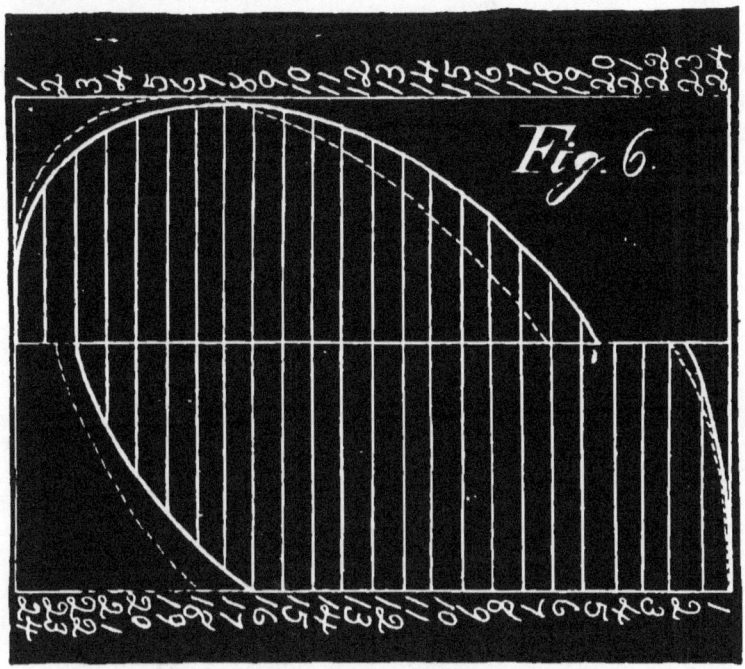

It will be observed, from this diagram, that the port nearest the crank obtained its maximum of opening as a steam port at the seventh inch of piston movement, but that it did not open to its full extent (the reason of this will be shown further on). It commenced to perceptibly reclose at the ninth inch of piston movement—the expansion began at $19\frac{1}{8}$ and ended at $22\frac{3}{4}$—and that the port opened as an exhaust port $\frac{7}{8}$ in. by the time the piston completed its stroke. As an exhaust port, it remained full open until the sixteenth-inch of piston movement, the cushioning at the 22d inch and the lead beginning to open close to the end of the stroke.

Now comparing the operation of this port to that of the other cylinder steam port (as denoted by the dotted diagram), we observe that it opened slower and remained open longer, it used the steam less expansively, the exhaust closed earlier, and the cushioning was greater. And furthermore, comparing the difference between the port openings during the two piston strokes with that shown in the diagram in Fig. 3, we find it a great deal more in this case.

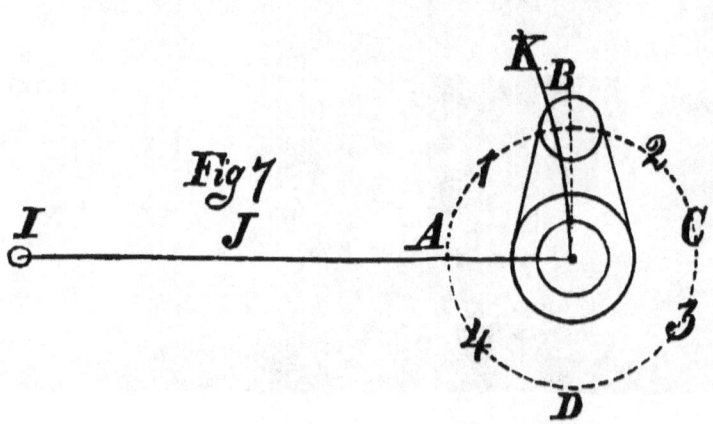

THE ANGULARITY OF THE CONNECTING ROD.

Suppose that in Fig. 7, the point, I, represents the diametrical center of a cross-head journal, and that the dotted circle represents the path of the center of the crank-pin. From A to B, then, will represent the first quarter-revolution of the crank (the figures from 1 to 4 denoting the respective quarter crank movements). The length of the line, J, from the center of the crank-shaft to the point, I, will represent the length of the connecting rod; and if, setting a pair of compasses to this length, and using the point, I, as a center, we mark the arc, K, it becomes evident that the point at the junction of the arc, K, with the dotted circle, is the point at which the center of the crank-pin will stand when the piston is in the middle of its stroke, and it follows that by the time that the crank has made a full quarter revolution (from A to B) the piston will have traveled more than half its stroke to an amount that is dependent upon the proportion existing between the length of the crank and that of the connecting rod, and the greater the relative length of the latter the less will that amount be. In the case under consideration, the stroke being 24 inches and the connecting rod 72 inches long, the piston will have traveled three quarters of an inch more than half of its stroke by the time that the crank has made its first quarter revolution, and it follows that during the second quarter crank-revolution the piston will travel less than half its stroke. For the third quarter crank revolution the piston movement will be the same in amount as it was during the second, and for the fourth the same as during the first.

The speed of the crank being uniform, it follows that that of the piston is irregular, and no matter in which direction the engine is running, the piston will travel relatively the fastest during each of the half-strokes performed between the end farthest from the crank and the middle of the cylinder. Suppose then that the opening and closure of the valve were performed uniform for each piston stroke, and that the crank made a half-revolution, starting from A in Fig. 7, traveling in the direction of the arrow and arriving at C. The speed of the piston being excessive during the first quarter crank movement, the opening permitted by the valve during that quarter revolution for the live steam to pass through is less in proportion to the piston movement than it should be. Now let us take the return stroke; while the crank is traveling from C to D, the piston will be moving below its average speed, and the port opening for the supply of steam being allowed as uniform for each stroke, the supply of steam will be greater in proportion to the piston movement.

If we turn to the diagram shown in Fig. 3, we shall find this variation shown very plainly, and yet very minutely; the egg-shaped diagram in full lines being the roundest on the right-hand side, which represents the end of the cylinder nearest to the crank. Confining ourselves to the half of the full line diagram above the line, $A\ B$, we find that the part of its length from 1 to $12\frac{3}{4}$, which represents the port-opening during the first quarter crank movement, we find it more oval-shaped than it is from $12\frac{3}{4}$ to 24, representing the second quarter crank movement, and the difference between the shapes of the two represents the precise difference arising from the varying piston movement, which is caused by the angularity of the connecting rod. Examining the effect of this angularity of rod upon the same port acting as an exhaust port, we find, from the part of the full line diagram below the line, $A\ B$, that the same irregularity exists, the only difference being that while it operated to diminish the opening of the port as a steam-port, it increased it as an exhaust port in a precisely similar ratio.

Turning now to the diagram shown in Fig. 6, we find that, although the relative lengths of the crank and connecting rod remained unchanged, yet the above conditions are reversed, for, whereas the diagram in Fig. 3 showed that the port (represented by the full line) acting as a steam port opened slower than it closed, that represented in Fig. 6 by the full lines (both representing the corresponding piston stroke) opened quicker than it closed. On the exhaust side (in both diagrams), however, it opened quicker than it closed, this effect being produced directly from the alterations made in the cylinder exhaust port—the valve—and the throw and lead of the eccentric, and influenced indirectly from varying the relations of these elements to the angularity of the connecting rod.

THE ECCENTRIC, PISTON AND CRANK MOVEMENTS.

The action of an eccentric movement converted into a reciprocating one is irregular, in the same proportion and for the same reasons already explained with reference to the crank and piston movement. In Fig. 8 is represented a crank and an eccentric, both being shown in position to commence the first quarter revolution, and it will be noted that while the angularity of the connecting rod will cause the piston to move, during the quarter crank revolution, at its quickest

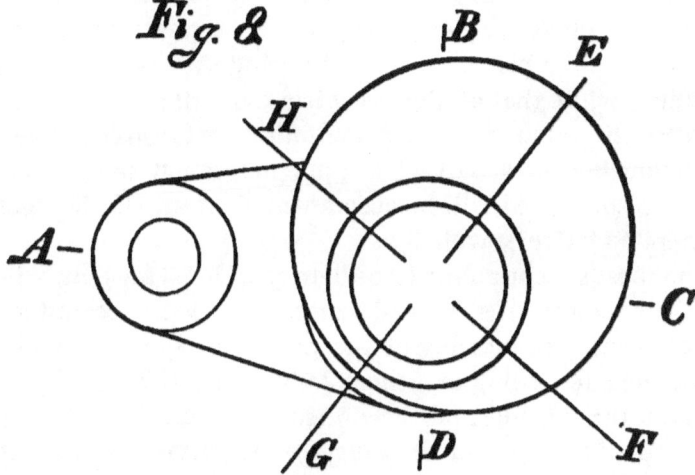

speed, the position of the eccentric is such that it will cause the angularity of the eccentric rod to diminish the valve travel during the greater part of the eccentric quarter revolution, for while the crank travels from A to B, in Fig. 8, the throw line of the eccentric will move from E to F. During this quarter revolution the valve is acting to open the steam-port (for the piston stroke under consideration) and hence the angularity of the eccentric rod is producing an effect the reverse of that due to the increase of valve travel, or in other words, the increase in the speed of the valve travel due to the increase of eccentric throw (necessitated by the use of the lap on the valve) is being modified by reason of the angularity of the eccentric rod.

During the second quarter crank revolution the angularity of the connecting rod will cause the piston to travel slower than during the first, while the angularity of eccentric rod will act to increase the valve travel during the greater part, and slightly diminish during the latter part, of the quarter revolution.

Turning now to the return piston stroke (the crank starting from the point C) the angularity of the connecting rod will act to retard the piston movement during the first quarter revolution, while that of the eccentric rod will act (mainly) to increase the valve travel (but to diminish it towards the last). While during the last quarter-crank movement the acceleration of piston speed will be accompanied by an (in the main) accelerated valve travel.

The most salient points to be here considered are that, when the piston is traveling fast and hence requires the greatest supply of steam, the opening of the port is the slowest, and *vice versa*, the effect being evidenced by the variation or distance between the full line and the dotted diagrams, in both Figs. 8 and 6. The object of drawing the diagrams one over the other, and distinguishing one of them by dotted lines, is for the especial purpose of making plainly visible and easily comparable the valve action, during one stroke as compared to the other, the effect due to the angularity of the connecting rod—the lap of the valve—and the travel of the valve being shown by the shape of the whole diagram on the left-hand side as compared to that of the right-hand side, and the effect of the lead of the eccentric being denoted by the variation between the shape of the full line and the dotted diagram.

In making a comparison, however, it must be remembered that the middle of the diagram is at the $12\frac{3}{4}$ division on the upper or steam side, and $11\frac{1}{4}$ division at the lower or exhaust side of the diagram, those points representing, respectively, the points at which the crank stands at a right angle to the connecting rod, or in other words, at half-throw.

It will be noted from the full line diagram in Fig. 6, that the port there represented never opened quite full as a steam port, and this always occurs when the valve is given lap, the deficiency increasing with the lap. It must also be remembered that during the opening of this port as a steam-port the piston speed is accelerated from the angularity of the connecting rod. To obviate this irregularity, the valve is often given more travel by increasing the throw of the eccentric; hence, examples with such increase will in due time be given.

It will also be seen, on referring to Fig. 6, that the expansion began at the $19\frac{3}{4}$ inch of piston movement and ended at the $22\frac{3}{4}$ inch in one case, while in the other stroke it began at the 18th inch and ended at $22\frac{1}{4}$ inch, and hence, that the steam was used expansively during three inches of one stroke and four and a quarter of the other. The exhaust also was unequal, taking place half an inch earlier for one stroke than for the other, and this distorted action always attends the employment of steam lap and increases in proportion to or with the lap, as will be shown in future examples of valve movements.

In further reviewing the port opening in relation to the piston travel, it must be borne in mind, that it is during the travel of the piston from the dead-center to about the third of its stroke, that a free port-opening is of the most importance, and more especially during the earlier part of the stroke, because it is then that the ports are opening, respectively, as a steam and an exhaust-port. It is for this reason that the lead of the valve is usually made equal for each stroke, irrespective of unequal points of exhaust.

On referring to Fig. 6, it will be seen that, while the ports opened quickly, their closure occurred slowly, the desirability of this action being self-evident, when we remember that after the piston has traveled, say, half its stroke under the full steam chest pressure, it would take considerable port-closure to materially affect the pressure in the cylinder between that point and the point of cut off, while on the other hand the exhaust port having opened quickly, and remained full open during two-thirds of the stroke, the greater part of the steam has escaped, and it would require considerable port-closure to effect any back pressure.

LAP ON THE EXHAUST SIDE OF THE VALVE.

When steam lap is employed to cut off the steam supply to the piston early in the stroke it opens the exhaust port earlier than is desirable, and to remedy this defect what is termed exhaust lap is sometimes added to the valve. The term exhaust lap means the amount to which the exhaust edges of the valve covers the exhaust edge of the cylinder steam port when in the middle of its travel. Thus, in Fig. 9, the dis-

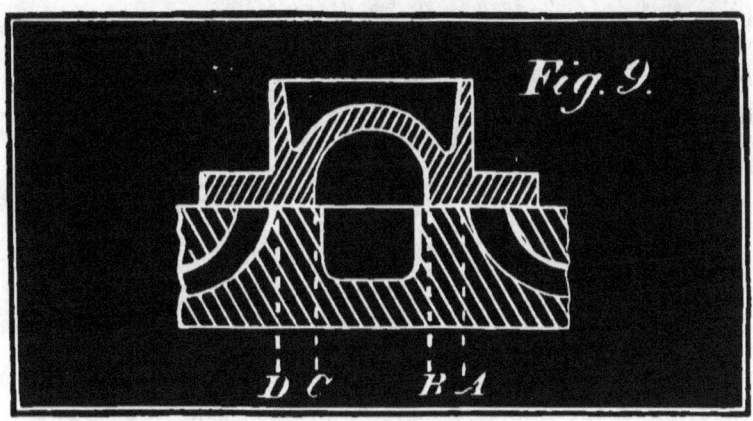

tance from A to B on the one side, and from C to D on the other side, is the exhaust lap. It is measured on one side only, so that from A to B and from C to D, measuring respectively ¾-inch, the valve would be said to have ¾-inch of exhaust lap.

The effect of exhaust lap is to enclose in the end of the cylinder a portion of the exhaust steam, thus compressing it and causing it to fill the clearance (that is the distance or space between the piston, when it is at the end of the stroke, and the cylinder cover) and the steam passage, at a pressure proportionate to the quantity of steam so enclosed, or, in other words, to the point of compression. The manner in which the exhaust lap performs this duty is by closing the exhaust-port before the piston has reached the end of the stroke. Thus, in Fig. 10, the valve is shown in the position in which it com-

mences this operation, and since the valve and piston are traveling in the direction denoted by the respective arrows, it follows that whatever steam remains in the steam passage, A, and in that end of the cylinder will be compressed by the advancing piston. This effect is sometimes termed the compression, and at others the cushion, one of its results being to cushion the piston; and in this respect, as well as in that of filling the clearance and the steam passage with steam at a pressure by the time the piston has arrived at the end of the stroke, it serves the same purpose as the valve lead, and that without drawing any steam from the steam chest.

Another effect of exhaust lap is to retain the live steam longer in the cylinder, or, in other words, to prolong the point of release. To offset these advantages, however, we have the serious disadvantage that it acts to diminish the exhaust area, and in view of this fact it becomes absolutely necessary, when exhaust lap is employed, to vary some other of the valve elements, in order to keep the exhaust free. In some cases the wid h of the cylinder exhaust-port is increased, in others the valve is given extreme travel as well. The introduction of these new elements creates so many considerations that it renders it much more difficult to calculate clearly the results, and renders the calculations much more liable to error. It is for these reasons that so much difference is found in actual practice, with reference to the employment of exhaust lap

In Fig. 11 is shown the valve from which the diagram in

Fig. 6 was taken, there having been (in Fig. 11) 5-16ths of exhaust lap added to it, and in Fig. 12 is shown a diagram of the valve movement under this new condition. From this diagram it will be observed that the exhaust is cramped early in the stroke, and (referring to the full line diagram) by adding up the total lengths of the vertical lines on the steam and

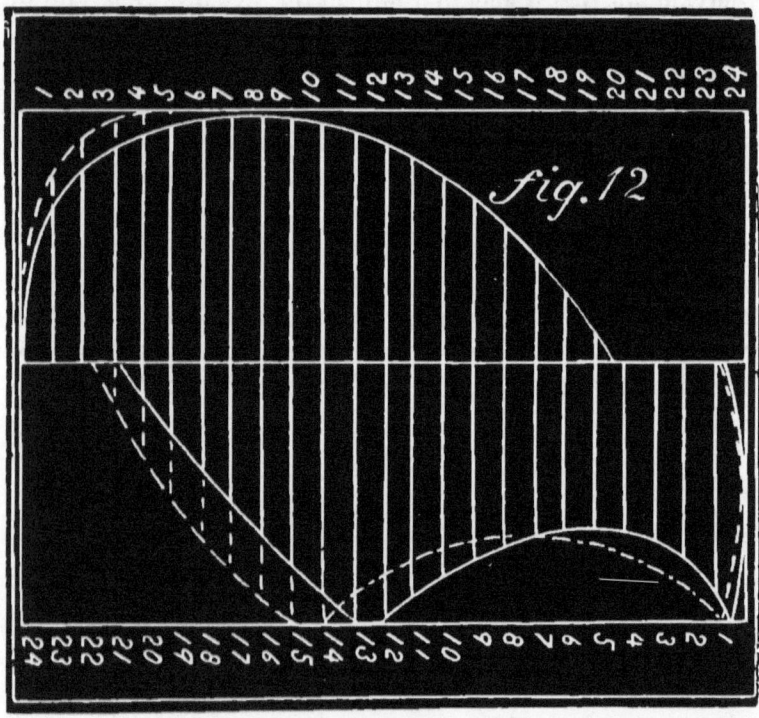

exhaust sides *respectively*, and then dividing their sum of the number of vertical lines (which will give us the average port openings) we shall find but a very little difference between the effective area of the port acting as a steam port and as an exhaust port, whereas the opening requires to be much greater for the exhaust than for the live steam, because, as the pressure of steam on the exhaust side is reduced, the exit of the steam remaining in the cylinder takes place slower, and hence, unless the piston movement is a very slow one, the exhaust lap will induce a back pressure on the exhaust side of the piston, which will far more than outweigh any advantage gained by the compression.

The partial exhaust closure, which is shown in Fig. 12 to to have taken place early in the exhaust, is due to the width at A, in Fig. 11, between the edge of the bridge and the exhaust edge of the valve, becoming narrowed by reason of the exhaust lap.

If the valve were given an increase of travel, the exhaust area would be still further diminished unless the cylinder exhaust-port were widened, in which event we have the cushion as the advantage, and a larger valve with increased travel as the cost.

In a large proportion of American, and in almost all English practice, exhaust lap is not employed, except it be from 1-32d to 1-16th-inch, or just sufficient to isolate the two steam ports from the valve exhaust port when the valve is in mid position.

CLEARANCE.

In Fig. 13 is shown a valve having what is termed clearance, which means the amount of opening between the edge of the cylinder port and the edge of valve exhaust-port when the valve is in the middle of its stroke, in which position it follows that there is open communication between both ends of the cylinder and the exhaust. The object sought by the employment of clearance is to free the exhaust, and its use is therefore mainly restricted to valves having but little steam lap, or having excessive travel, and upon fast running engines.

Were we to give clearance to a valve having no steam lap, the result would be that when it opened to admit steam, the latter would pass clear through the same port to the valve exhaust-port, until such time as the valve had opened on the steam side, to an amount equal to that of the clearance on the valve, and it follows, therefore, that the clearance must always be less than the steam lap.

In Fig. 14 is shown the action of a valve having $\frac{1}{4}$-inch steam lap and 3-16th inch of clearance. The movement is shown for one stroke only, so as to keep the lines clear of one another.

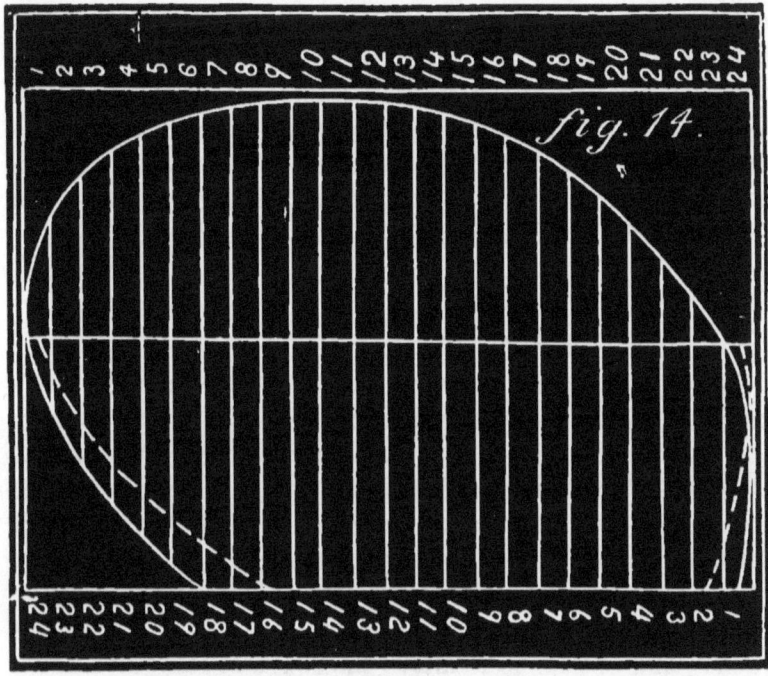

The full line represents the valve action with the clearance, and the dotted lines show how the exhaust would take place if the valve had no clearance. The valve, it will be noted, begins to open ⅜-inch earlier and is opened full an inch earlier, remaining so during 2½ inches more of piston stroke, than it would without the clearance. It also takes away the compression.

If this valve was used upon an engine running very slowly, it is a fine point as to whether the clearance would in this case, be desirable or not, but if the engine ran at an ordinary speed there can be no question but that the clearance would be highly beneficial.

It is evident that to give clearance to the valve whose movement is shown in Fig. 5, would be useless because the exhaust is in that case open $\frac{7}{8}$-inch by the time the piston arrives at the end of the stroke—is full open when the piston has traveled 1-16th inch of the return stroke and remains open during 22 inches of the exhaust stroke ; and the only effect of adding clearance would be to open the steam-port earlier as an exhaust-port, and thus permit the steam to escape too soon.

VALVE TRAVEL.

We have in all our previous examples given to the valve a travel equal to twice the width of steam-port added to twice the amount of steam lap, which accords more with English than with American practice, and we may, therefore, proceed to examine the effect of an increase of valve travel (which is common in American practice), first applying it to the valve shown in Fig. 6, whose elements we will leave the same as before, save that the travel is increased from $3\frac{1}{2}$ inches to 4 inches, and in Fig. 15 is shown the action of the valve under this new condition.

The effect of the alteration has been that whereas the ports, acting as steam ports, obtained their maximum opening at the seventh inch for one, and at the fifth inch for the other piston movement, they were opened under the increased travel at the 2d and $2\frac{3}{4}$ respective inches of piston movement. On the exhaust side, however, a partial reclosure of the valve has taken place early in the stroke, while the cushion has been reduced $\frac{3}{8}$ inch for one stroke, and increased $\frac{5}{8}$ inch for the other.

The average exhaust opening has been reduced in consequence of the partial reclosure of the port, the area of ex-

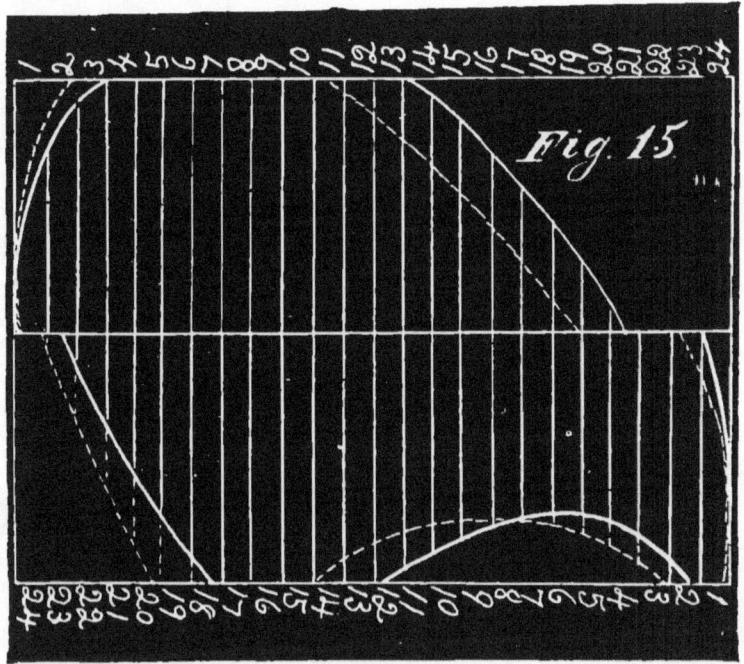

haust opening becoming but very little greater than that for the live steam, and this, upon the general principle that the port opening for the exhaust requires to be considerably greater, and that in the rules for obtaining the required port area this fact is recognized, does not appear to be any advantage, since we have filched from the exhaust to add to the steam-port opening. Our alteration of valve travel has, furthermore, reduced the expansion for one stroke, so that instead of lasting during 3 inches, it lasts during $2\frac{5}{8}$ inches of piston movement, while for the other stroke the alteration has been to decrease the amount of expansion from occurring during $4\frac{1}{4}$ inches to $3\frac{1}{2}$ inches of piston movement.

The release, as the point at which the exhaust begins is

termed, has, under the increased travel, taken place about ¾ inch later for one stroke, and ¼ inch later for the other stroke, and this under ordinary circumstances would be of decided advantage, but in view, in this case, of the diminished area of exhaust opening, it becomes of but little if any advantage.

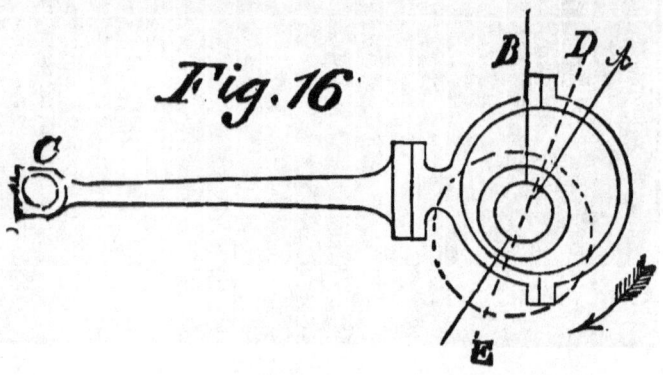

In order to understand clearly the reason why the increase of valve travel has made so much difference to the valve movement, it becomes necessary to examine into the alteration which has been made in the position of the eccentric. Suppose then, that in Fig. 16, A represents the throw line of the eccentric, and B a line at a right angle to a line drawn from the center of the eccentric rod eye, C, to the center of the bore of the eccentric, then from A to B will represent the lead of the eccentric, and it follows that, since the distance between A and B increases in proportion as the points at which they are measured is remote from the center of the eccentric bore, therefore, the longer the throw of the eccentric the more is its lead' notwithstanding that the throw line of the eccentric remains at the same number of degrees of angle to the line, B, so that by increasing the throw of the eccentric, we have rendered it necessary to take off some of its angular advance or lead, as denoted in Fig. 16, in which the dotted line, D, represents the position of the throw line of the eccentric when adjusted to suit the increase of travel. If now we bear in mind that the throw lines are shown in the position they would stand in when the piston was at the end of the stroke, we shall perceive as follows :—

It is self-evident that when the throw line of the eccentric stands in the position of the right angle line B in Fig. 16, the eccentric will impart a quicker motion to the valve than it will at any subsequent part of its next half revolution in the direction of the arrow, and it follows that when the piston is at the end of the stroke, and the steam-port is open to the amount of the lead, the nearer the throw line of the eccentric stands to the right angle line, B, the quicker will be the movement of the valve while opening the steam-port.

On referring to the dotted circle and line, E, in Fig. 16, which represent the position of the eccentric when the piston is at the other end of the stroke, it will be apparent that the same remarks hold good, hence, as we increase the valve travel, or, what is the same thing, the throw of the eccentric, we must decrease the amount of eccentric lead in order to have the same amount of valve lead, and this alteration causes the opening of the ports to take place more quickly.

In order to remove the partial exhaust closure, shown in Fig. 15 to have occurred during the early part of the exhaust, we may adopt either of the following methods :—

The closure occurred from a contraction of the width at A, in Fig. 11. The amount of the closure, we find by measuring it in Fig. 15, was 9-32d inch, by narrowing the bridges to that amount, and thus making the cylinder exhaust-port 9-16th wider the partial closure would be removed. This, however, would leave the bridges 15-32d inch wide only, and would reduce their capability to resist the abrasion due to the friction of the valve face. Were this alteration made, however, the amount of contact between the valve face when the valve was at the end of its travel and the face of the bridge, would be ¼ inch in width, which is sufficient to form a steam-tight joint.

Another plan would be to make each bridge 9-32d inch wider; in this case, however, a new valve made 9-16th inch wider in its exhaust-port would be reqnired. This plan increases the wearing qualifications of the bridges.

A third method would be to leave the bridges the original width (¾ inch) and to make the cylinder exhaust-port 9-16th inch wider, and to make the valve exhaust-port wider to the same amount.

It is obvious that the plan first mentioned would be applicable for altering existing ports, so as to save getting a new valve.

POINT OF CUT-OFF.

If the elements of a slide valve motion be designed to cut off the steam supply to the piston at a point earlier in the stroke than when the piston has travelled about three-quarters of its stroke, the action of the valve will be very irregular for one piston stroke as compared to the other, and for this reason the slide valve is not employed upon engines requiring to have the steam supply cut off earlier than at most two-thirds of the stroke, unless, indeed, the engine have a link motion which serves to diminish the valve travel and thus make the point of cut-off earlier.

In Fig. 17 there is illustrated the action of a valve having the following elements :—Width of steam ports, 1 in. ; width of bridges, $\frac{1}{2}$ in. ; width of cylinder exhaust port, 2 in. ; steam lap, $1\frac{1}{8}$ in. ; exhaust lap, 1-64th in. ; travel, $4\frac{1}{4}$ in. ; lead of valve, $\frac{1}{8}$ in. ; piston stroke, 26 in. ; length of connecting rod, 72 in.

It will be noted that, whereas in all our previous examples the width of the bridges has been $\frac{3}{4}$ inch, it is now made an inch. This has been done so that (the cylinder exhaust-port being twice the width of the steam-port, and the valve travel equal to twice the width of steam-port added to twice the amount of the lap) the valve shall move so as to open the ports

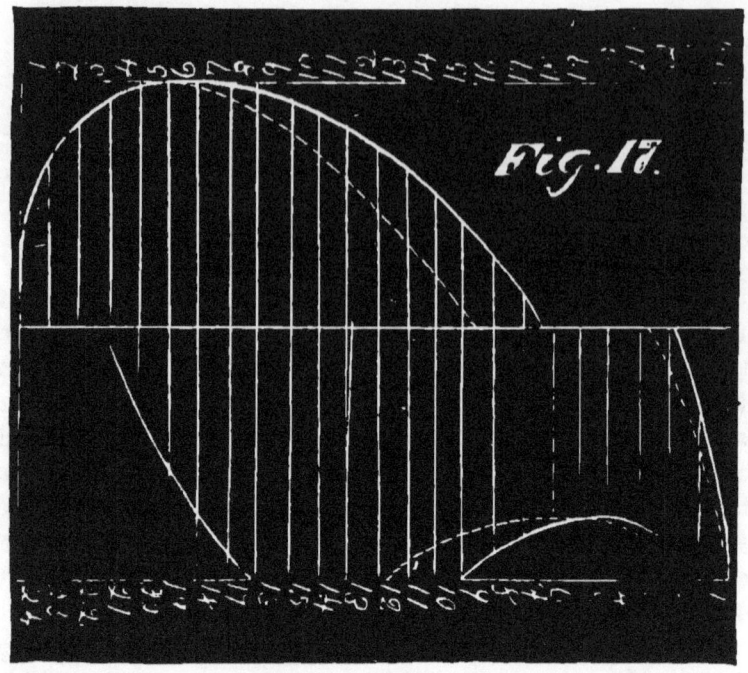

full as steam-ports, which would not be the case with narrower bridges.

In Fig. 17 is represented the action of the above valve, and the first thing to attract our attention is that the exhaust has taken place when the piston had arrived at $21\frac{3}{8}$ inches for one stroke and at $22\frac{1}{4}$ for the other, thus giving very early points of release.

The expansion commenced for one stroke at the $17\frac{3}{4}$ inch of piston stroke and lasted during $4\frac{1}{2}$ inches of the stroke, and for the other, at the $15\frac{1}{2}$ inch, and lasted during $5\frac{7}{8}$ inches of the stroke, or, in other words, the point of cut-off occurred at nearly two-thirds of one stroke, and at nearly three-quarters of the other stroke.

Comparing the area of port opening of the exhaust with that of the steam, we find, referring to the stroke represented by the full lines, that the area of that port as an exhaust was a trifle greater than it was as a steam-port, whereas on the other stroke, as shown by the dotted lines, the area of the port opening was considerably greater than it was acting as a steam-port. The cushion also occurs during more than an inch more of one stroke than of the other.

The travel of the valve we find has been just sufficient to open the steam-ports full, and any greater travel would proportionately diminish the exhaust area by increasing the partial closure at the cylinder exhaust-port shown in Fig. 17, on the full line diagram, to have taken place from nearly the first to the ninth inch of piston movement.

To prolong the points of release the valve would require the addition of exhaust lap. Such addition, however, would diminish the exhaust area and thus prove undesirable. It would also increase the cushion. Any clearance given to the valve exhaust-port would act to make the points of release earlier, whereas they occurred too early as it is.

It is only left then to accommodate an increase of valve travel by increasing either the width of the bridges or the width of the cylinder exhaust-port, and since the adoption of either plan necessitates to an equal extent the widening of the valve, it matters little which plan is taken. The objection to enlarging the valve is that its friction to its seat, and hence the power necessary to operate it is increased. It follows that by variations in the travel, width of bridges, width of cylinder exhaust-port, lap on steam and exhaust side, and clearance on the valve, an almost endless variety of valve movements may be designed, and are, in fact, found in practice; but while it is quite praticable to the engineering student to ascertain by calculation the action of any given valve and ports, yet dry

4*

figures do not impress the memory, especially when each individual movement possesses so many points to be remembered. By calling in the aid of the eye and showing the whole of the valve action in a diagram, the results of any given combination of elements are so much more easily grasped and retained that it becomes desirable to illustrate the action of the valves on various existing engines by the same kind of diagram as we have used in explaining the functions of the various valve elements.

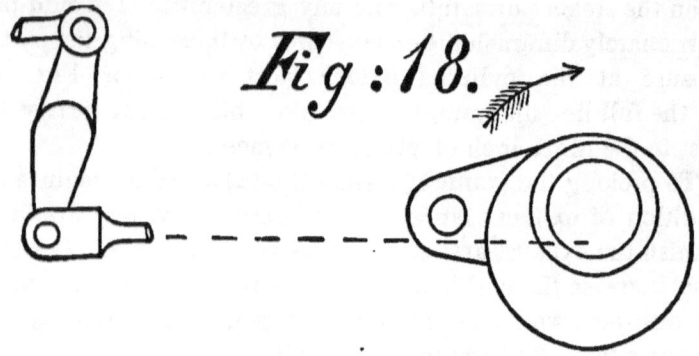

Fig: 18.

THE ROCK-SHAFT.

If the eccentric rod connects to a rock-shaft, as shown in Fig. 18, the position of the eccentric requires to be altered, because the rock-shaft reverses the motion of the valve with relation to that of the eccentric; and furthermore, the latter will require a less amount of angular advance or lead, to maintain the same amount of valve lead, as we shall presently perceive. By introducing the rock-shaft we also reduce the length of the eccentric rod.

In Fig. 19 are shown diagrams of a valve movement having the same elements as those shown in Fig. 6, save that in this case a rock-shaft having arms 5½ inches long is introduced; the eccentric rod being shortened from 4 feet 10 inches (the length during our previous experiments) to 3 feet 5 inches long.

Here it will be observed that, whereas in all our former diagrams the full line figure (representing the stroke during which the piston travels towards the crank) has shown the ports for both the steam and exhaust to open the slowest, the introduction of the rock-shaft has caused them to open the quickest during that piston stroke, and since the piston is travelling the fastest during the first half of this stroke, the steam port opening becomes more proportionate to the piston movement.

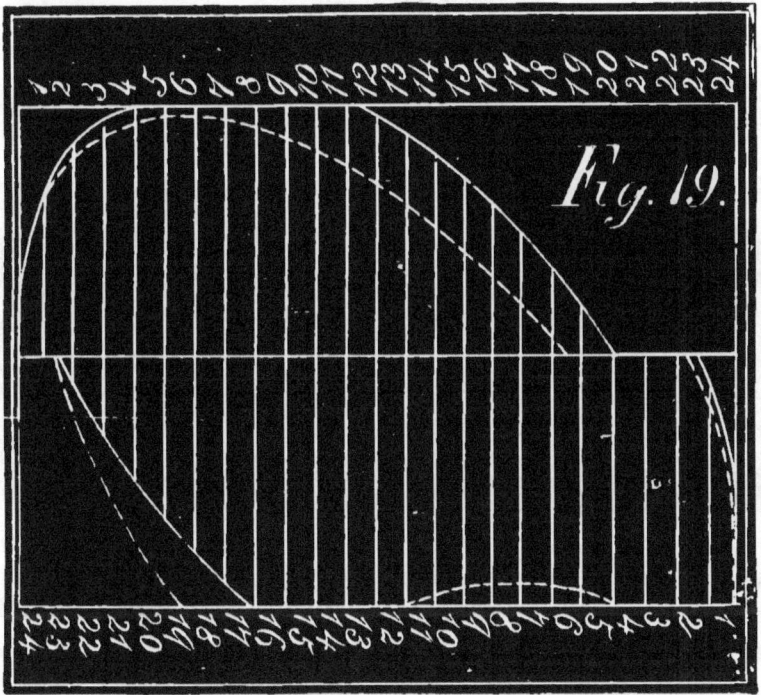

Our diagram (Fig. 19) shows the ports acting as steam ports, to have opened three inches earlier for one, and one inch (of piston movement) earlier for the other piston movement than was the case in Fig. 6, and the reason is as follows:—

Suppose that in Fig. 20, A represents the center of a crank-shaft, and that the circle B represents the path of the center of the eccentric. Suppose also the engine to have no rock-shaft, the crank-pin being at S, the direction of rotation being as denoted by the arrow. If, then, the eccentric be given an angular advance of 30°, its throw-line will stand at E.

Now suppose that P represents the steam edge of one steam port and R represents the steam edge of the other steam port, both being equidistant from the centre H of the ports, then the amount of valve lead due to the eccentric advance from C to E (30°) will be the distance between G and R (the length of the eccentric rod being from A to H, which measures the same as line K or the line from E to G).

Fig. 20.

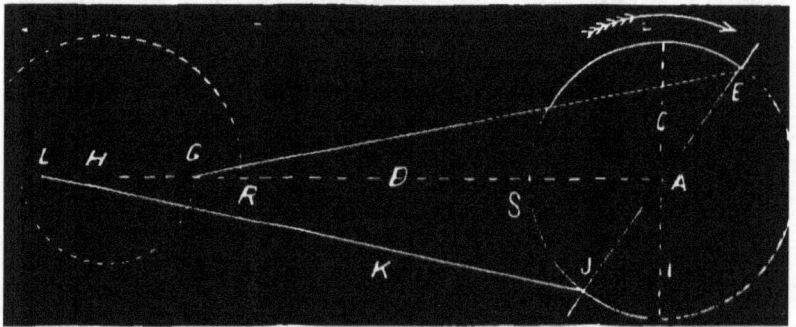

Now suppose the engine to have a rockshaft, and the eccentric (moved to its required position) be given the same 30° of angular advance, as at J, then the amount of valve lead due to this angular advance is the distance from P to L, which is greater than that from G to R. Hence it follows that with a given amount of eccentric angular advance the amount of port lead will be greater in an engine having a rockshaft, or what is the same thing. For a given amount of valve lead less angular advance is required for the eccentric in the one position than would be required were it in the other. Now the nearer the throw line of the eccentric stands to a line at a right angle to the crank, the faster the valve movement while the crank is at and near the dead centers, and hence the more rapid port openings.

By turning the eccentric upon the shaft we have reversed the effect of the irregularity of valve travel (due to the angularity of the eccentric rod) so that when the angularity of the connecting rod is (starting from a dead center) accelerating the piston movement, the angularity of the eccentric rod is also accelerating the valve movement. When, however, the piston starts from the end of the cylinder nearest to the crank, and the angularity of the connecting rod is retarding the piston movement, the angularity of the eccentric rod will be acting to accelerate the valve movement, and hence is is that in Fig. 19 the full line diagram assumes the chracteristics of the dotted ones in former examples, by opening both the earliest and the fullest.

EXAMPLES OF VALVE MOVEMENTS, TAKEN FROM THE MOST RECENT LOCOMOTIVE PRACTICE.

The following examples show the action of the various valves when under full travel, and, therefore, irrespective of the action of the link motions when employed to reduce the valve travel so as to use the steam more expansively.

Fig. 21 represents the action of a valve having the same elements as those upon an ordinary 16×24 inch cylinder American passenger locomotive, the elements being as follows: Length of steam-ports, 15 in.; width of steam-ports, $1\frac{1}{4}$ in.; width of bridges, 1 in.; width of cylinder exhaust port, $2\frac{1}{2}$ in.; outside (or steam) lap, $\frac{3}{4}$ in.; inside (or exhaust) lap 1-32d in.; travel of valve $5\frac{3}{8}$ in.

The full line diagram represents the stroke when the piston is traveling away from the crank, the engine running forward.

In Fig. 22 is shown similar diagrams of an English engine, the design of James Cudworth, the ports, bridges, etc., being of the same dimensions as the above, the only point of difference being that the Cudworth engine has $\frac{7}{8}$ inch of steam-lap, and the valve travel is $4\frac{1}{2}$ inches only.

The most prominent difference between these respective

diagrams are as follows: The English engine used the steam expansively during 1¾ inch more of one, and 1⅜ inch of the other piston stroke, and has a much more free exhaust. Its point of release, however, took place an inch earlier in one, and 11-16th inch earlier in the other stroke, *per contra*,

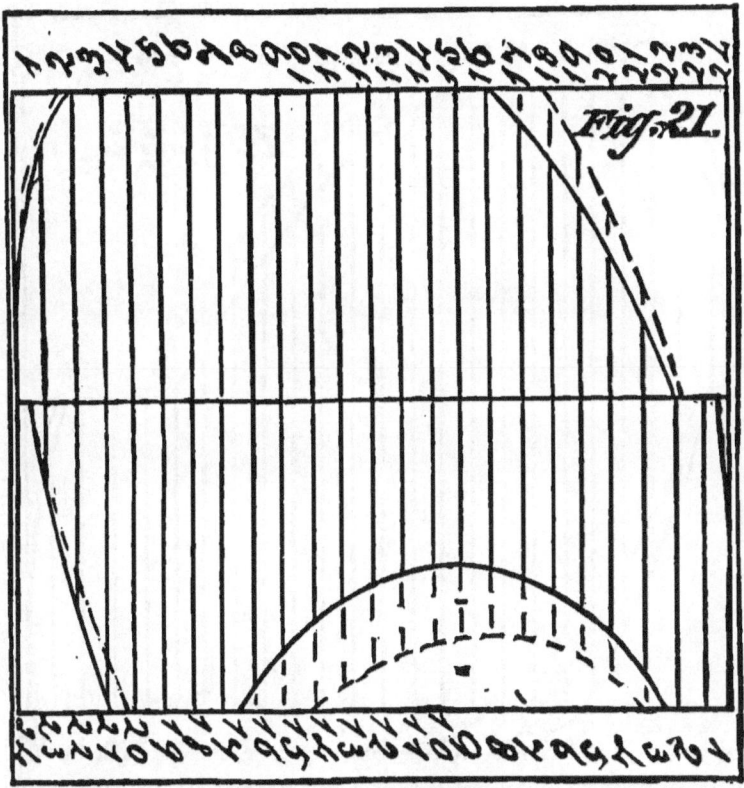

however, it cushioned on the exhaust ⅜ inch more for one stroke, and an inch more for the other stroke. In prolonging the point of release the Baldwin engine possesses an advantage that is of vastly more consequence than is the earlier exhaust cushion, and the main other consideration is the exhaust area Let us take the other two full line diagrams and compare them by adding the length of the lines and dividing the sum by the number of lines, which will give us the average respective port openings, and we shall find as follows: Average width of steam-port opening of the Baldwin engine, from

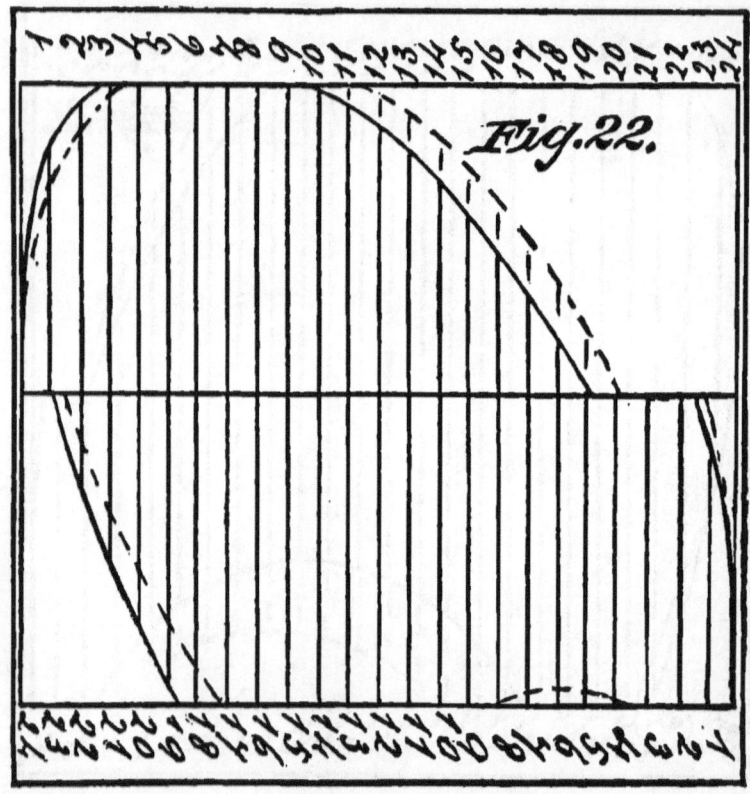

Fig. 22.

the beginning of the stroke to the point of cut-off, 1.07 inch. Average exhaust opening of the Baldwin engine from the opening to the closure of exhaust 15-16th inch.

Average width of steam-port opening of English engine from the beginning of stroke to the point of cut-off, 9-10th inch. Average exhaust opening of English engine from beginning of exhaust to closure of same 1.12 inch.

It must also be observed that the loss of exhaust area in the American engine commences from the $2\frac{1}{2}$ inch of the exhaust stroke and continues to the $16\frac{1}{2}$ inch, and that, although in the English engine the port acting as a steam port commenced its closure 6 inches earlier and continued it during over 3 inches more of the stroke, this is of little import, since the port being full open until the piston had traveled $9\frac{3}{4}$ inches of the stroke, and the port area being proportioned, as is generally presumed to be the case in view of its acting alternately as a steam and exhaust, and, therefore, larger than it would be required if it acted as a steam-port, only a slight variation in the reclosure is but of little moment, and the same remarks will apply to the slight difference in the opening of the port as a steam-port. Thus we then have, when the valve is given full stroke, (that is to say independently of cutting off earlier in the stroke by the employment the link motion) the broad issue of the advantage of prolonging the point of release an inch at the cost of reducing the exhaust area as above noted, this issue being involved in the increase in the valve travel.

In Fig. 23 we have a diagram of the port openings of an engine built for the Midland (English) Railroad, the dimensions being as follows:

Steam-ports $1\frac{1}{4}$ inches by 15 inches; cylinder exhaust port $3\frac{1}{2}$ inches; bridge between ports 1 inch; steam lap 1 inch; exhaust lap 0; travel of valve $4\frac{1}{2}$ inches; lead $\frac{1}{8}$ inch; size of cylinder 17×24 inches; length of connecting rod 5 feet 11 inches. In consequence of the shortness of the valve travel we find here that the ports do not open to their full width as steam ports, and as a result, while the average steam port opening is $\frac{3}{4}$ inch, the average exhaust port opening is $1\frac{1}{10}$

inches. The point of release is at the 22d for one and the 23d inch for the other stroke; while to partly offset this early release, the cushion amounts to about $1\frac{1}{2}$ inches for one and two inches for the other stroke. Another result of the small valve travel is that the expansion takes place during $4\frac{1}{2}$ inches of one and 4 inches of the other stroke, which is during a large portion of the stroke when compared to valves having a maximun of travel. The admission of the steam is identical for the two strokes during the first five inches of the piston movement. A peculiar feature of this valve is the width of the cylinder exhaust port, which might be narrowed one inch without affecting the diagram in the slightest particular; the size of the valve might then be reduced an inch, with a corresponding diminution in the power required to operate it. In the design of this valve freedom of exhaust has evidently been the main consideration.

In Fig. 24, we have a diagram of the port openings of a Rogers Locomotive Works (of Paterson, N. J.) 16×24 engine, the dimensions being as follows:

Steam-ports $1\frac{3}{16}\times15$ inches; cylinder exhaust port $2\frac{1}{2}$ in-

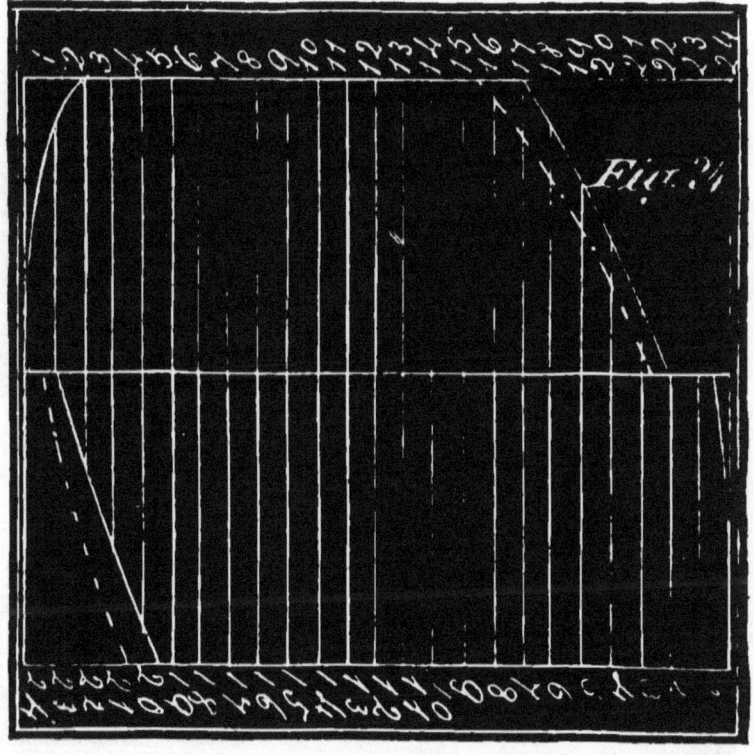

ches; bridge between cylinder ports 1¼ inches; outside lap ¾ inch; inside lap ⅛ inch; travel of valve 5 inches; lead of valve ⅛ inches; length of rock shaft arm 9 inches; length of connecting rod 6 feet 10½ inches.

A notable feature in this diagram is the large exhaust area, notwithstanding the over travel of the valve, and also of there being ⅛ inch of inside lap, and this is due to the extra width of the bridges between the cylinder ports. The average steam-port effective area is 1.01 inches, while the average effective area of exhaust port is 1.07 inches, the latter being greater

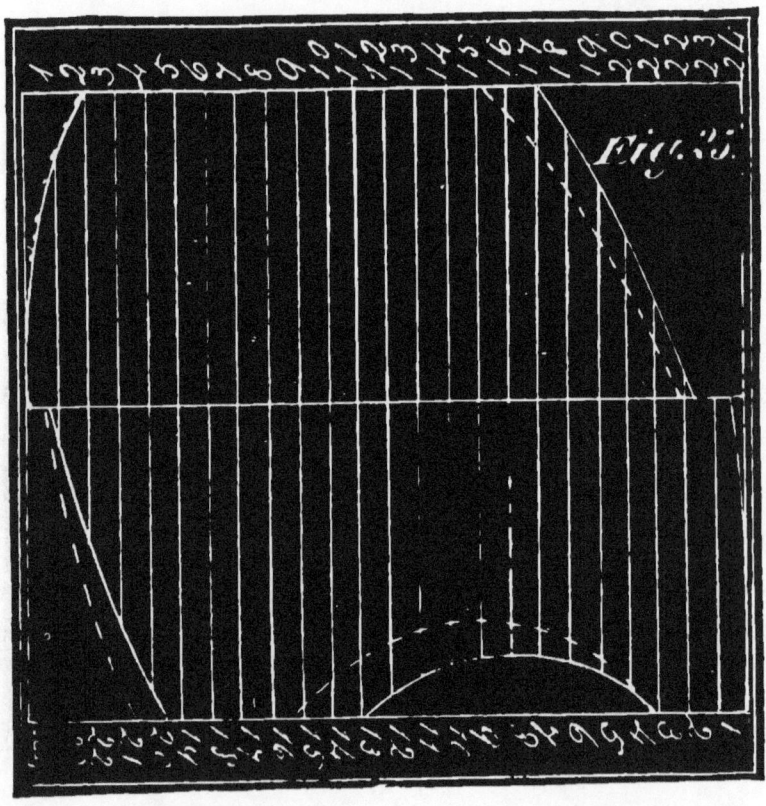

than that shown in Fig. 21, notwithstanding that the ports are $\frac{1}{16}$ inch narrower (the outside lap being the same in both cases).

In Fig. 25 are shown the port openings of a 16×24 inch cylinder, the design of the Grant Locomotive Works (of Paterson, N. J.) the dimensions being as follows:

Steam-ports $1\frac{1}{4}$×14 inches; cylinder exhaust port $2\frac{1}{2}$ inches; bridge between ports 1 inch; outside lap $\frac{3}{4}$ inch; inside 0 inch; travel of valve 5 inches; lead of valve $\frac{1}{32}$ inch; length of rocker arm $10\frac{1}{4}$ inches; length of connecting rod 7 feet 1 inch.

In this diagram the points of admission, release and cushion are unusually regular for one stroke as compared to the other. It will be seen that in all those valves having increase travel, the steam is used less expansively, but the point of cut off and release are very even, whereas in those valves having lesser travel the expansion is more variable, the points of release more variable, and the cushion greater.

In Fig. 26 is shown the port openings of an engine, the proportions of whose valve details represent the average for American freight locomotives, the elements being as follows:

Steam ports $1\frac{1}{4}$×14 inches; width of bridge between ports 1 inch; outside lap $\frac{3}{4}$ inch; inside lap $\frac{3}{8}$ inch; lead $\frac{1}{10}$ inch; valve travel 5 inches; length of connecting rod 7 feet; length of rocker arm 10 inches.

The average steam-port opening is here shown to be 1.09 inches, and the average exhaust opening is $\frac{8}{10}$ inch. The reclosure of the exhaust port commences it will be observed, when the piston has moved but $1\frac{1}{2}$ inch of the stroke, and while acting as a steam-port, the ports remained open during some 13 inches of the piston stroke; when acting as exhaust ports they remained open during about $2\frac{1}{2}$ inches of piston stroke.

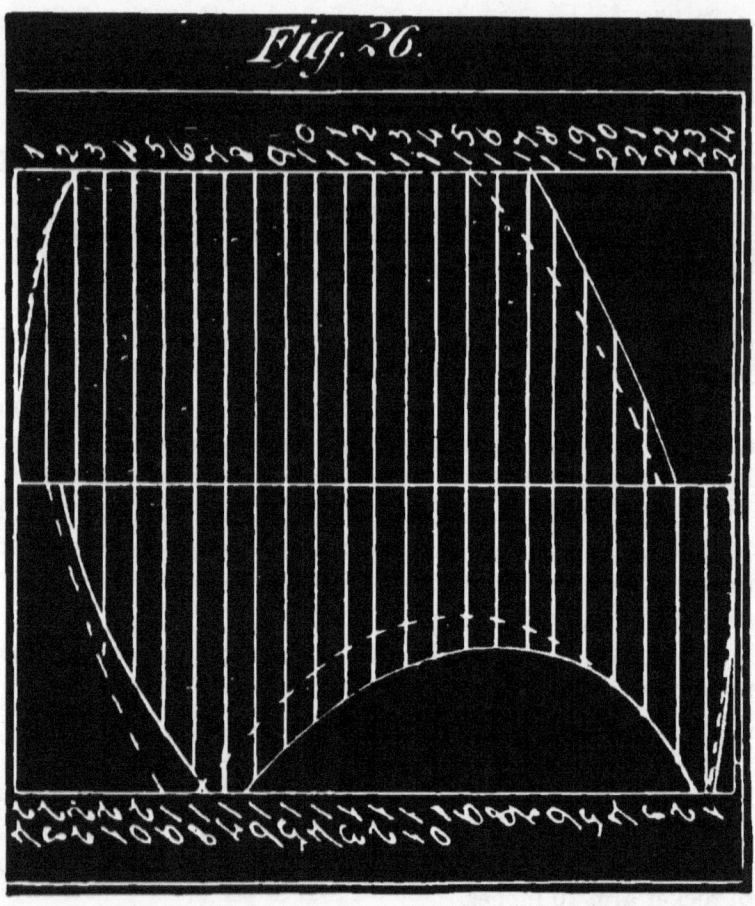

In Fig. 27 is a diagram of the port openings of a representative of the engines used upon the Susquehanna division of the Erie R. R., the proportions being as follows:

Size of cylinders 18×24 inches; width of steam ports 1 inch; length of steam ports 16 inches; width of cylinder exhaust port 2¼ inches; width of bridge between ports 1⅜ inch; outside lap 1¾⁄₁₆ inch; inside lap 3⁄32 inch; travel of valve 5 inches;

lead of valve $\frac{1}{16}$ inch; length of connecting rod 7 feet 2 inches.

In these proportions the extreme length as compared to the width of the steam-ports and the increased width of the bridge are notable features. The average opening of the steam-port is $\frac{8}{10}$ inch and that of the exhaust port is 1.05 inch.

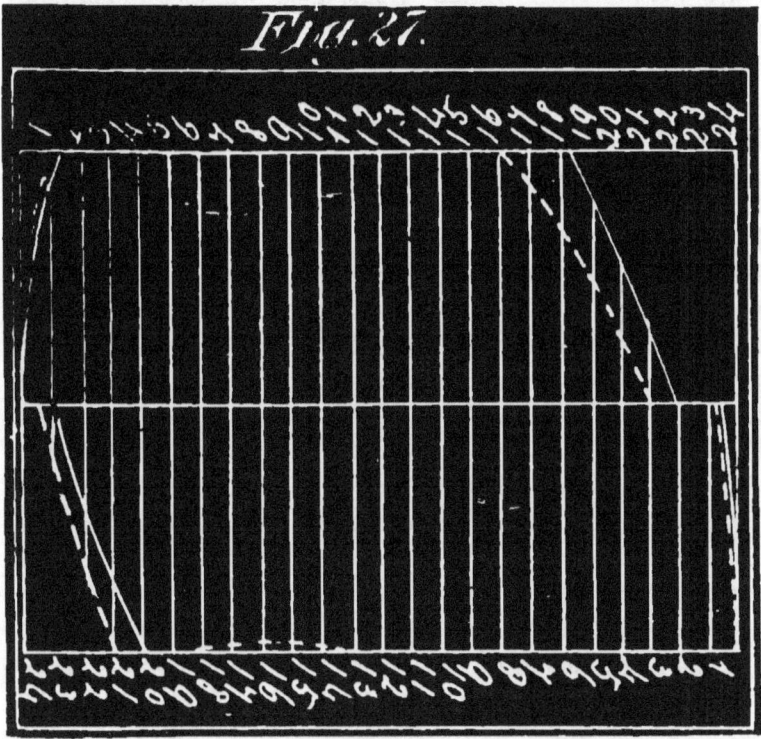

We may now institute a comparison of the actual admission and exhaust area of the ports whose openings are illustrated in the various diagrams, for it will be noted that the average amount of openings so far given have referred to the width of opening irrespective of the lengths of the ports.

First, however, it may be pointed out that the element of time is an important consideration, for a given area of steam port which would be ample under a slow speed may be altogether inadequate under a much increased speed of engine. Indeed it is obvious that with sufficient time a comparatively very small port area will give ample admission and exhaust. As a practical illustration of this fact, we have the circumstance that a great many locomotives will run faster with the links hooked up to shorten the valve travel, than with the valve under full travel. The reason of this is that with the valve in full travel the steam follows the piston during a greater portion of the stroke, and hence there is more steam to exhaust during the exhaust stroke, and there is not time, in such case, for the exhaust to be sufficiently free to avoid back pressure. That back pressure must in such cases exist is proven by the following considerations. The amount of the engine lead being a given quantity, it is obvious that with a free exhaust the further the live steam follows the piston the faster the engine should run, or to use other terms, if, with the live steam following the piston (say) two inches more of the stroke in one case than in another, and the engine speed is the same in both cases (the amount of the load remaining constant,) where are we to look for the duty due to the increased consumption of steam in the cylinder during those two inches of stroke, and for the increased pressure (due to the same two inches of live steam,) which exists during that part of the stroke in which the steam was working expansively? The amount of power represented by that steam has been lost to the boiler and cannot be found in the duty except it be that the exhaust pressure is counted as duty. Possibly in many cases it may not be altogether lost for the following reasons:

If the exhaust is cramped there may be left in the cylinder a large portion of the exhaust steam, which is compressed and used with the live steam during the next stroke; supposing this then to be the case, we have an engine large enough in all its parts to do a certain amount of duty, but from a contraction of exhaust area the engine can never be employed to perform that amount of duty save at being started. This exception indeed deserves particular attention as applied to locomotives, the reasoning being as follows :

In starting, the speed being slow, there is time, with the valve in full travel, for the steam to exhaust without undue loss from back pressure, and when the train is under headway the travel of the valve may be reduced by hooking up the link. Again, when ascending a grade sufficiently steep to reduce the engine speed, the valve may be given full travel. Thus full travel may only be used when the speed of the engine being reduced there is given time for the steam to find a practically free exit. The considerations then resolve themselves into this. Is it more desirable that the full power of an engine shall be capable of being exerted at starting, and under heavy duty due to steep grades, or during the average duty during the entire trip? and furthermore, what are the objections to giving to the cylinder an exhaust practically free at the greatest speed the engine may be called upon to run? If there are many steep grades and frequent stoppages, the first proposition will readily commend itself, unless the latter proves unattainable by reason of involving other and equivalent disadvantages. If, however, we carefully investigate the diagrams we shall scarcely find such to be the case, in many of them at least, for those diagrams in which no reclosure of the exhaust port has been shown to take place, and in which therefore the effective area of the ports, acting as exhaust ports, exceeds their effective areas when acting as steam-ports, will bear favorable comparison with those in which such is not the case. It is worthy of note also, that the reclosure of the exhaust port usually occurs from over valve travel or from exhaust lap; when over travel is the cause the steam is used less expansively in the cylinder and at the same time the cushioning is less; on the other hand, however, the points of admission and cut off are more nearly equal in the two strokes. When exhaust lap is the cause, the expansion and the cushion are increased at the expense of the exhaust.

Here however another consideration as applied to locomotives claims attention, inasmuch as if we consider the train speed, the size of the driving wheels makes a difference. Suppose for instance that an engine is traveling at 20 miles an hour, the driving wheels being 4 feet 6 inches in diameter, a greater number of piston strokes will be made in a given time than would be the case if the engine ran at the same speed, and the wheels were 5 feet 6 inches in diameter; hence, with the valve and port elements of the same proportions in both engines, the exhaust would, in the latter case, be more free, because it would have more time in which to take place.

THE RESULTS OF THE FOREGOING EXAMPLES COMPARED.

The great difference in the action of a valve shown (by our diagrams of examples) to accompany apparently slight differences in the elements composing the valve mechanism, and the numerous ways in which variations may be made, render it difficult to compare the value of one valve movement with that of another, or with others, as the case may be. Hence it becomes necessary to so present a summary of the data given by the diagrams, that the results of all the examples in any particular of their operation may be read without referring to the diagrams. This we may best accomplish by means of a series of tables, the result obtained from each table being given in its last column, hence to proceed. Let the first column in table number 1 represent the number of the diagram referred to; the second column the size of the steam-ports; the third the size of the cylinder, and the fourth the distance the live steam followed the piston, or point of cut off, and the fifth the cubical contents of the space in the cylinder which requires to be fil.ed with live steam at each stroke.

TABLE 1.

Fig.	Size of Ports. Inches.	Size of cyl. Inches.	Point of cut off. Inches.	Cubic feet of live steam per stroke.
21	1¼ x 15	16 x 24	22	*2·56
22	1¼ x 15	16 x 24	19½	2·26
23	1¼ x 15	16 x 24	18½	2·15
24	1₁₀³ x 15	16 x 24	22	2·56
25	1¼ x 14	16 x 24	22	2·56
26	1¼ x 14	16 x 24	22	2·56
27	1 x 16	18 x 24	22	3·26

We may now compose another table as follows: The first column shall refer to the number of the diagram as before; the second shall contain the average *width* of steam-port opening (as shown by the respective diagrams); the third shall contain the effective area of the steam-port (obtained by multiplying the average width of steam-port opening by the length of the port); the fourth shall contain the volume of live steam required per piston stroke, and the fifth shall present the proportion of the port area to the live steam volume. Thus we shall have the space requiring to be filled at each stroke with live steam, and the effective area each port affords through which to fill it, while the last column will afford means of comparing the values of the ports relative to the duty.

TABLE 2.

Fig.	Average width of steam-port opening. Inches.	Effective steam-port area. Square inches.		Volume of live steam. Cubic feet.		Sq. inches of port area per foot of live steam.
21	1·07	†16·05	÷	2·56	=	6·270
22	·9	13·50	÷	2·26	=	5·973
23	·75	11·25	÷	2·15	=	5·232
24	1·01	15·15	÷	2·56	=	5·918
25	1·01	14·14	÷	2·56	=	5·523
26	1·09	15·26	÷	2·56	=	5·960
27	1·05	16·80	÷	3·26	=	5·153

* This column is obtained by multiplying the area of the cylinder by the point of cut off, the question of clearance being omitted for simplification.

†This column is the product of the second column in this table multiplied by the length of port given in the second column of our previous table.

Here, then, we find that the greatest proportion of steam-port area as compared to the volume of steam to be admittted is given by the valve whose movement is shown in Fig. 21, while that shown in Fig. 22 is second, and Fig. 26 stands third, and this notwithstanding that average width of port opening is less in numbers 21 and 22 than in number 26. The reason of this is that 21 and 26 have a maximum of width of steam-port opening, while number 22 has a shorter period of admission. In the English engines (Figs. 22 and 23) the points of cut off take place earlier, and the amount of expansion is greater, hence the effective steam-port area is large as compared to the volume of steam to be admitted. In the next table the first column remains the same; the second column represents the number of inches of the piston stroke during which the steam was used expansively. The third column represents the proportion of the stroke during which the piston was propelled by the steam (the full length of the stroke being represented by 48), and the fourth column gives the amount of piston movement during which the compression occurred.

TABLE 3.

Fig.	Amount of expansion in inches of stroke.	Steam followed piston.	Amount of compression inches of stroke.
21	$1\frac{1}{2}$	$\frac{47}{48}$ of whole stroke.	$\frac{1}{2}$
22	3	$\frac{45}{48}$ " "	$1\frac{1}{4}$
23	$4\frac{1}{4}$	$\frac{45}{48}$ " "	$1\frac{3}{4}$
24	$1\frac{3}{4}$	$\frac{47}{48}$ " "	$\frac{3}{4}$
25	$1\frac{5}{8}$	$\frac{47}{48}$ " "	$\frac{5}{8}$
26	2	$\frac{47}{48}$ " "	$1\frac{1}{4}$
27	$1\frac{7}{8}$	$\frac{47}{48}$ " "	$\frac{3}{4}$

It remains now to examine the relative volumes of steam contained in the respective steam passages, which we may do as in Table number 4, in which however the length of the steam passage is obtained as follows: The question of clearance between the piston and the cylinder cover (that is the space between the cylinder cover and the piston when the latter is at the end of the stroke) is left out, since by supposing all the engines to have the same amount of clearance, that item need not be considered. Now the length of all the piston strokes is 24 inches, and the half of this is 12 from this latter number we deduct the width of one bridge added to half the width of the cylinder exhaust port, and the remainder is given as the length of the steam passage. This assumes the distance of the steam chest face from the cylinder bore to be equal in all the engines, an assumption which enables us to keep our tables clear of complication by disregarding the trivial or exceptional differences in this respect which may perhaps be found in practice

The steam contained in the passage, however, gives out expansive power from the point of cut off to the point of release, or in other words during the term of expansion, hence it becomes necessary to credit each engine with that item, which is done in the column under the heading of "expansive part" of stroke.

TABLE 4.

Fig.	Port area Sq. inches x	Length of passage. Lineal ins. =	Contents of passage. Cubic ins.	Expansive part of stroke. Lineal inches.
21	18·75	9·75	182·81	1½
22	18·75	9·75	182·81	3
23	18·75	9·25	173·43	4¼
24	17·80	9·50	169·10	1¾
25	17·50	9·75	170·62	1⅝
26	17·50	9·75	172·62	2
27	16·00	9·50	152·52	1⅞

From this we find that in number 21 less power is imparted to the piston by the live steam contained in the steam passages than is the case in any other, while 23 stands first and 22 next in order of superiority. To present the relative effectiveness of the ports in a more simple form we make a table as below, in which the second column contains the comparative value of the port area in proportion to the cubic feet of live steam to be discharged, taking the port having the greatest area as being 1. Then in the third column we may place the comparative value of the expansive force obtained from steam in the steam passages, the maximum (obtained by Fig. 23) being taken as 1.

TABLE 5.

Fig.	Comparative value of port areas per cubic foot of live steam.	Comparative value obtained from steam in passages.
21	1·00	·35
22	·95	·70
23	·82	1·00
24	·94	·41
25	·88	·38
26	·95	·47
27	·82	·44

Turning now to the same ports acting as exhaust ports we find as follows:

TABLE 6.

Fig.	Average width of exhaust openings. Lineal inches.		Length of ports. Inches.		Average effective exhaust area. Sq. inches
21	·937	×	15	=	14·05
22	1·120		15		16·80
23	1·110		15		16·50
24	1·070		15		16·05
25	1·070		14		15·98
26	·800		14		11·20
27	1·050		16		16·80

From the result obtained in the last column of the above table we may compose another as follows:

TABLE 7.

Fig.	Average effective exhaust area. Sq. inches.	Volume of steam to be discharged per stroke. Cub c feet.	Sq. inches of exhaust area per cubic foot of exhaust steam.
21	14·05 ÷	*2·665 =	5·27
22	16·80	2·365	7·10
23	16·50	2·250	7·33
24	16·05	2·652	6·05
25	15·98	2·658	6·01
26	11·20	2·659	4·21
27	16·80	3·348	5·01

*The volume here given is the quantity of live steam admitted by the port acting as a steam port, or, in other words, the volume of live steam admitted to the cylinder and steam passage.

Taking the exhaust area of Fig. 23 (which bears the largest proportion to its duty) as being represented by 1, we shall have the following relative exhaust port values.

TABLE 8.

Fig.	Relative value of exhaust port to its duty.
23	1·00
22	·96
24	·82
25	·82
21	·72
27	·68
26	·57

Here then we find a very great disproportion to exist between the exhaust port areas and the duty, and the number of piston strokes, or wheel revolutions, in a given time being taken as equal for all, it is self-evident that the ports having the highest rates are unnecessarily large, and therefore wasteful of steam, or else those having the lowest, being too small in effective area, produce back pressure.

If, however, in the case of a locomotive, the train speed instead of the revolutions were taken, the size of the driving wheels and the speed at which the engine is designed to run would require to be taken into account and would alter the result. It may be noted, however, that the committee appointed by the Railroad Master Mechanics' Association, in a recent report upon the subject of the sizes of locomotive cylinder ports, preferred to take the revolutions as we have done in our considerations. It may also be noted that the ports of Figs. 22 and 23 are English and fast running engines, requiring therefore larger exhaust areas; on the other hand, however, the sizes of their driving wheels are larger. As applied to stationary engines our deductions cover the whole ground because the revolutions represent the engine speed.

Another and important consideration with reference to these exhaust areas is as follows. When the exhaust area is diminished in consequence of the partial reclosure of the cylinder exhaust port, as shown in Figs. 21, 25 and 26, the question of time assumes a new importance, inasmuch as there may be a considerable back pressure during the early part of the exhaust stroke which cannot be compensated for by any amount of increased exhaust freedom during the latter part of that stroke. An examination of the diagrams will show the ports acting as exhaust ports to have closed, as shown in the table below.

TABLE 9.

Figs.	Began to close at inches of stroke.	Finally closed at inches of stroke.
21	20½	23½
22	18	22¾
23	16½	22⅝
24	20	23¼
25	20	23⅜
26		
27		

Hence the ports shown in Figs. 21, 25 and 26 remained

6*

opened longer than those shown in Figs. 22 and 23, but the latter remained full open up to the point where the closure began, while the former had partly reclosed early in the stroke. To estimate, therefore, the relative values of the exhaust we must take this fact into consideration, which may be done as follows:

By taking the piston strokes as being equal in a given time we are enabled to form an estimate of the values of the respective ports and valve proportions in relation to their respective cylinders and duties; and for this purpose we may assume the wheels to be of equal diameters. Now suppose that a point on the periphery of the respective driving wheels moves at the rate of the 48th part of a revolution in two seconds; now the exhaust takes place during one half of a revolution of the driving wheel, hence we may reason as follows: In Fig. 28 the half circle represents the half revolution

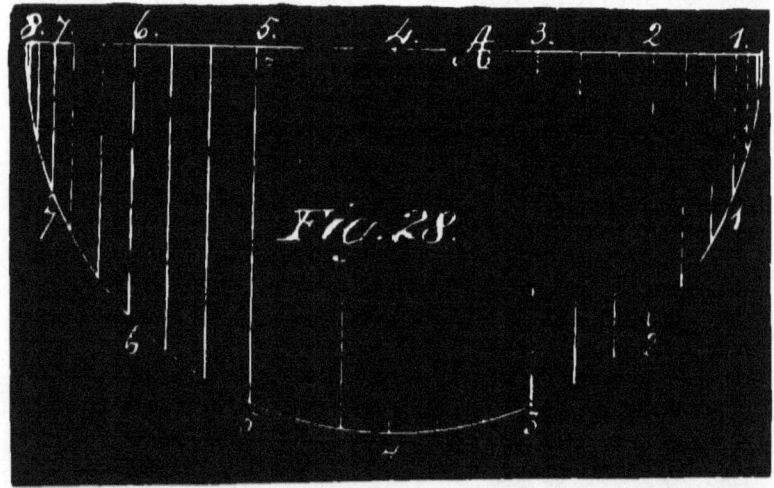

during which the exhaust occurs, and the 24 divisions upon its circumference represent each the twenty-fourth part of its half revolution. It will be apparent then that the time occupied by a point on the periphery of the driving wheel will be one second, in which to travel from one division to the next, or, what is the same thing, three seconds to travel three divisions, and as it will be easier and more explicit to take three seconds than one, we will divide off the half revolution into eight periods, each representing three seconds of time. Now it is apparent that while the driving wheel is traveling through either of these respective periods the piston will move through a distance equal to the distance contained in each of the several periods measured along the hoizontal line A, as denoted by the respective figures and marks above that line. Now it will be noted that we have marked off the half circle into as many equal divisions as there are inches in the piston strokes, hence, while those on the half circle will represent equal periods of time for the motion of the fly-wheel, those on the horizontal line will denote inches of piston movement during that time. We have only then to compare the average time and piston movement (which will be equal for all the engines) in each of the eight divisions with the average port areas shown in the diagrams for each division, and we shall have made a comparison of the effectiveness of the exhaust ports with the element of time taken into account.

TABLE 10.
*AVERAGE EXHAUST PORT AREAS DURING PERIODS.

PERIODS.	Fig. 21 Inc.	Fig. 22 Inches.	Fig. 23 Inches.	Fig. 24 Inches.	Fig. 25 Inches.	Fig. 26 Inches	Fig. 27 Inches
1st 3 seconds	17·8	18·75	18·75	18·75	17·50	15·12	16·00
2d "	12·6	18·60	18·75	18·75	15·12	9·94	16·00
3d "	10·3	18·60	18·75	18·75	13·10	8·68	16·00
4th "	10·3	18·75	18·75	18·75	14·56	9·52	16·00
5th "	13·1	18·75	18·75	18·75	16·52	11·76	16·00
6th "	14·4	18·75	16·20	18·75	17·50	16·80	16·00
7th "	18·1	10·80	9·05	15·60	15·54	12·74	15·20
8th "	6·6	1,80	·18	2·25	2·80	·21	4·00

*The above average areas are obtained by multiplying the average lengths of the three respective lines on the exhaust side of each diagram (representing the respective period) by the length of the steam-ports.

To carry the data contained in the above table a step farther we might take the respective areas for each period, and calculate the quantity of steam that would escape through during the period, and from that deduce the pressure on the exhaust side of the piston when the same had reached the part of the stroke represented by the end of each period; this however is scarcely necessary to our purpose.

It will be instructive however to deduce from the foregoing tables the relative areas of the ports acting as steam and as exhaust ports, and this we may do in the following manner. In table number 11, the second column contains the volume of live steam requiring to pass through the port to fill the cylinder (up to the point of cut off), and the steam passage. The third column contains the effective area of the port acting as a steam port, and the fourth column shows the ratio of effective exhaust area to the effective steam area.

TABLE 11.

Fig.	*Volume of steam passing through steam-port Cubic feet.	Average effective steam-port area. Square inches.	Average effective exhaust area. Square inches.	Ratio of exhaust area to steam area.
23	2·250	11·25	16·50	1·47
27	3·348	12·80	16·80	1·31
22	2·365	13·50	16·80	1·24
21	2·665	16·05	14·05	1·14
25	2·658	14·14	15·98	1·13
24	2·652	15·15	16·05	1·06
26	2·659	15·26	11·20	·73

*This volume is the total requiring to be admitted and exhausted.

Here then we find that whereas the exhaust area in Fig. 23 is .47 greater than the admission area in Fig. 26, it is ·27 less. Again, while the exhaust area, in Fig. 24 is ·06 larger than the admission, in Fig. 27, it is ·31 greater, and the consideration naturally arises that if in Fig. 24 the exhaust is sufficiently free the admission is unnecessarily large, hence the valve travel might be reduced sufficiently to give no more than the requisite admission area, in which case the steam would be used more expansively and the exhaust area would be increased while the valve would require less power to operate it. In this connection it may be noted that a committee of the Railroad Master Mechanics have recommended an addition of steam lap over that usually given, in order to secure more expansion; now since a reduction of valve travel will operate in the same direction, as well as effecting the above stated advantages, it appears in every way desirable. Supposing, however, that, as shown in Fig. 27, an exhaust ·37 larger than the admission area is necessary to avoid undue back pressure, then the valve travels of Figs. 21, 25 and 26 may be reduced, increasing their exhaust areas, using the steam more expansively and reducing the power necessary to their operation, as well as the wear of the valve and seat faces.

The wearing surfaces of the respective valves are given in table number 12, the valves being allowed to overlap the ports an inch at each end of the ports. The wearing surfaces of the seats are given, so far as concerns the bridges, between the cylinder ports only, because the width of seat face outside of the steam-ports varies considerably in practice. The surface of the bridges, however, always wears faster than the surface outside the ports, and furthermore the bridge surface always wears hollow in its length, as well as hollow as denoted by a straight-edge placed across the cylinder face and at a right angle to the length of the port.

Table 12.

Fig.	Sizes of valves. Inches.	Area of valve. Sq. inches.	Area of bridges. Sq. inches.	Valve travel. Inches.
21	7½ x 17	127·5	15	5⅜
22	7¾ x 17	131·7	15	4½
23	10 x 17	170·0	15	4½
24	8⅞ x 17	150·8	15	5
25	7½ x 16	120·0	14	5
26	7½ x 16	120·0	14	5
27	8⅜ x 18	145·2	16	5

AREA OF STEAM PORTS, SHAPE OF STEAM PORTS, AND WIDTH OF BRIDGES.

The author has thought it the best to give the rules for finding the area of steam ports as given by recognized authorities on that subject, and to supplement these with numerous examples taken from general practice, and the reader will at once observe the wide variation existing between theory and practice on this subject. It will be noted that the practice of extensive locomotive builders has been selected, and it may be further remarked that in stationary steam engine practice the same variation exists. It is necessary to add, however, that in comparing the area of the steam port with that of the steam cylinder only, the piston speed is assumed to be equal in each case, whereas, if much discrepancy existed in the piston speed, and the smallest ports accompanied the slowest piston speed, their relative efficiency would be increased because of the increase of time during which the steam would flow through them for each piston stroke.

The area of a steam port should bear a definite proportion to the quantity of steam required to pass through it in a *given time,* and the following are the rules given by the various authorities named.

Mr. Bourne, in his Catechism of the Steam Engine, says, "In slow-working engines the common size of the cylinder passages is one-twenty-fifth of the area of the cylinder, or one-fifth the diameter of the cylinder, which is the same thing. This proportion corresponds very nearly with one square inch per horse-power when the length of the cylinder is about equal to its diameter ; and one square inch of area per horse-power for the cylinder ports and eduction passages answers very well in the case of engines working at the ordinary piston speed of 220 feet per minute." The same author also says, "The area of the ports of locomotive engines is usually so proportioned as to be from one-tenth to one-eighth the area of the cylinder, in some cases even as much as one-sixth, and in all high speed engines the ports should be very large and the valve should have a good deal of travel, so as to open the port very quickly."

Dr. Zeuner, in his work on the slide valve, gives the following : "To find the proper area of port for an engine of a given piston speed, multiply the area of the piston in square inches by the number opposite to the given piston speed on the table below.

Speed of piston in feet per minute.	Number whereby to multiply the area of the piston.
100	0·02
200	0·04
300	0·06
400	0·07
500	0·09
600	0·1
700	0·12
800	0·14
900	0·15
1000	0·17

In "The Slide Valve Practically Considered," by N. P. Burgh, is the following : "The author's extended experience in engines of all classes, embracing all the modern improvements by the best makers, enables him to give the following correct formula for the area of the opening or port caused by the valve for the supply steam :—

High pressure engines — HP. × ·6 to ·5 square inches. In "Link and Valve Motions," by Auchincloss, it says, referring to the area of steam ports: "This dimension ranks next to cut-off in its controlling influence upon the proportions of the valve seat and face. It may justly be considered as a *base* from which all the other dimensions are derived in conformity with certain laws. Its value depends greatly upon the manner in which the port is employed, whether simply for admitting the steam to the cylinder, or for purposes both of admission and exit. In cases of admission it is evident that the pressure will be sustained at substantially a constant quantity by the flow of steam from the boiler. But in case of exit or exhaust, a limited quantity of steam, impelled by a constantly diminishing pressure, forces its way into the atmosphere with less and less velocity. If, then, the engine is supplied with two steam and two exhaust passages, the ports will be correctly proportioned when the areas of the latter exceed those of the former by an amount indicated by careful experiment. When, however, one passage performs both duties, it should have an area suitable for the exhaust, and be opened only a limited amount for the admission of steam. Very excellent results have been found to attend the employment of an area equal to 0·04 of that of the piston, and a steam pipe area of 0·025 of the same, when the speed of the piston does not exceed 200 feet per minute; but widely different factors are demanded by higher speeds like those peculiar to locomotives.

In the year 1846 MM. Gouin and Le Chatelier instituted a series of experiments for ascertaining the value of such terms. These were continued about six years later by Messrs. Clark, Gooch, and Bertera upon engines of British manufacture. The various results having been collated and analyzed by Mr. Clark, were finally presented to the public in his valuable work on "Railway Locomotives." From this it appears that, with a piston speed of 600 feet per minute, an area of 0·1 that of the piston was found to give practically a perfect exhaust."

The same author then gives the following table for intermediate speeds of piston "on the assumption that a higher speed is usually attended by increased pressure."

Speed of piston.	Port area.
200 feet per minute.	·04 area of piston.
250 " "	·047 " "
300 " "	·055 " "
350 " "	·062 " "
400 " "	·07 " "
450 " "	·077 " "
500 " "	·085 " "
550 " "	·092 " "
600 " "	·1 " "

In practice, however, the proportions vary considerably in some cases, because of peculiar requirements, and in others without any very apparent reason; especially is this latter the case in the matter of locomotives, as the following table discloses:

Name of Engine.	Size of cylinder.	Size of steam ports.	Area of steam ports
	Inches.	Inches.	Inches.
The Duke (English)	18×26	1½×15 =	22½
Baldwin Engine,	18×24	1¼×16 =	20
Cudworth Engine (English)	17×24	1¼×15	18¾
P. & R. Railway,	18×24	1¼×15	18¾
Dickson Mfg. Co. Engine,	17×24	1¼×16	20
"Adirondack" B. & A.R.R.	18×26	1¼×10	12½
"Brown." B. & A. R.R.	18×26	1⅛×14	15¾

Here it will be seen that the two last named engines have
the largest cylinders and the smallest port area. They are
the freight engines which have lately been undergoing tests
upon the Boston and Albany railroad. In this connection,
however, it may be noted that the apparent difference is some-
what modified by variations in the valve travel—the first engine named in the list having but $3\frac{7}{8}$ inches valve travel,
while the two last named have 5 inches.

We next come to a consideration of the proportions desirable between the width and length of a port. Suppose, for example, its area is determined upon as $18\frac{3}{4}$ inches. It may be
$1 \times 18\frac{3}{4}$ inches, $1\frac{1}{8} \times 16{\cdot}66$ inches, $1\frac{1}{4} \times 15$ inches, or $1\frac{1}{2} \times 12{\cdot}5$
inches; the area and contents of the passages remaining the
same. It is self-evident that the length of the port must bear
some proportion to the diameter of the cylinder, though our
table shows considerable latitude in this respect. A long and
narrow port gives the advantage that when the piston is at
the commencement of its stroke, and the valve has left the
steam port, say $\frac{1}{4}$ inch open, the latter, if 18 inches long,
would permit the steam to pass into the cylinder through an
area of $18 \times \frac{1}{4} = 4\frac{1}{2}$ inches; whereas with a $1\frac{1}{2} \times 12\frac{1}{2}$-inch port
standing $\frac{1}{4}$ inch open, there would be but an area of $12\frac{1}{2} \times 1\frac{1}{2}$
—$3\frac{1}{8}$ inches, only of opening for the steam to pass through, so
that the long and narrow port would give a more ready supply of steam to the piston during the earlier portion of the
stroke, assisting the exhaust also, to a similar degree; furthermore, the stroke of the valve is reduced. On the other hand,
however, the length of the valve is increased, and its width
diminished, and it becomes a question as to whether the long,
narrow, and short stroke valve will require more or less power
to operate than will the other. Suppose, then, that the area
of port required is $18\frac{3}{4}$ inches, and that we examine into the
respective merits of a port $18\frac{3}{4} \times 1$ inch, and one $12\frac{1}{2} \times 1\frac{1}{2}$ inch.
Allowing, in each case, the bridges to be an inch wide, the

cylinder exhaust port to be twice the width of the steam port, the valves covering the ports an inch at each end, the valve having (as in Fig. 1) no lap, and just sufficient travel to leave the steam ports full open. The sizes of the respective valves will be $20\frac{3}{4} \times 6$ inches for the long port, and $14\frac{1}{2} \times 8$ inches for the short port; the respective areas being $124\frac{1}{2}$ inches for the first, and 116 inches for the last named. In considering the pressure with which these respectives valves are pressed to their seats, and hence the power required to operate them, we may disregard (for simplicity's sake) the counteracting pressure placed by the live and exhaust steam on the underneath side of the valve, since that part of the subject will be treated of in connection with valves having steam lap, becau e it then becomes a more important element. To proceed then, we multiply these respective areas by the steam pressure under which they are supposed to operate (say 100 lbs.), we shall obtain the total force with which the valves are pressed to their seats. We may now, for the sake of illustration, presume that each 100 lbs. of pressure of the valve to its seat will require a force of 10 lbs. to move the valve, and we have as follows:—

Total pressure of valve to seat.	Equivalent of friction.	Force required to move valve.	Travel of valve.	Power required to move valve one stroke.
Lbs.	Lbs.	Lbs.	Inches.	Inch Lbs.
For long port, 12450	÷ 10 =	1245	2	2490
For short " 11600	÷ 10 =	1160	3	3480

Here then it appears that the valve for the long port will require about 44 per cent. of the power to move it that is required by its opponent; and since it has been shown to give the most ready supply of steam, it is in both respects the most desirable. There would, however, be more friction of the steam upon the sides of the passages, and probably a greater amount of condensation would take place.

In this connection it may also be observed, that in the case of the long port the steam passages would be longer to an amount equal to the difference in width between the two ports compared ; these, however, are minor elements, especially the latter, providing that the cut-off takes place early in the stroke, in which event the steam in the passage performs expansive duty after the cut-off takes place.

WIDTH OF BRIDGES.

We next come to a consideration of the width of the bridges between the cylinder steam and exhaust ports (shown in Fig. 1, at D D). The proper width of the bridge is often given as the same in amount as the thickness of the cylinder ; there is no reason, however, why this width should not be proportioned to its duty. It has first to separate the steam port from the exhaust port, and must therefore be strong enough to withstand the highest pressure (with the usual margin of extra strength) to which the cylinder is intended to be subjected. Here, however, it may be noted that the thickness of the bridge need not be uniform, but may be cut away at and near the face of the seat, thus obtaining a strong bridge with a narrow seat face. It must be remembered that a saving of a quarter of an inch in the width of the bridge, makes half an inch less in the required width of the valve, and by reducing its area, operates (in the same manner as the narrow port) to diminish the amount of power necessary to operate the valve.

It has, secondly, to maintain a steam-tight joint (with the valve) between the cylinder steam and exhaust ports. To form such a joint, a quarter of an inch is sufficient, so that if, at the extreme ends of the valve travel, there is left that width of contact between the valve and seat faces, a tight joint may be assured ; but whether so narrow a bridge as this would give will stand the wear so as to maintain a tight joint, is another question, and one depending upon several elements. If the valve travel is greater than that necessary to permit to the port a full opening, or in other words, if that travel is greater in amount than twice the width of the steam port in a valve (such as shown in Fig. 1), or twice the width of the steam port, added to twice the amount of any steam lap that valve may have added to it, then the bridge has more duty placed upon it, in consequence of the necessary addition to the travel, and it should therefore have an addition of area to withstand the resulting additional wear.

The hardness of the metal of which the seat is composed, and the pressure of steam under which the valve is required to operate, are also elements which go to determine the wearing qualifications of the bridge. It is evident, however, that (presuming the iron to be as hard as it can conveniently be operated upon by the machine cutting-tools) the area of the bridge should bear a definite proportion to the pressure of the valve to its seat, and the amount of the valve travel, or in other words, it should bear a definite relation to its duty; the proportion of that relation being determined by results taken from actual experiment.

The width of bridge is generally less than the width of the steam port, varying in English locomotive practice from about one and three-eighths to one and a quarter, and in American from fifteen-sixteenths to one and a quarter inch, and the bridges are found in both cases to usually wear hollow in their lengths.

THE POWER REQUIRED TO OPERATE A SLIDE VALVE.

Any mathematical rule whereby to calculate the power required to operate a slide valve is vitiated by the following facts: The rule must be based upon the unbalanced area of the valve, the force or weight pressing the valve to its seat, and the distance the valve moves through in a given time. The first and last of these elements may be very easily determined, but the force pressing the valve to its seat is a quantity which varies with the shape of the valve, its size, strength, and the temperature under which it operates, or, in other words, it varies with the fit of the valve to its seat, and that fit is in a continuous state of change.

When a valve is lying upon its seat in the open air, there should be, if it is fitted to its seat, the atmospheric pressure of say 15 lbs. per square inch of unbalanced area pressing it to its seat. The unbalanced area will be composed of the area of valve face in contact with the iron surface of the steam chest face or valve seat face, and will amount, in a small valve, to, say 25 inches, which, multiplied by the atmospheric pressure, say 15 lbs., would give 375 lbs. in addition to the weight of the valve as the power necessary to lift the valve from its seat. As a practical fact, however, it does not matter how perfectly any valve, when new, may fit to its seat; one month's wear will so entirely destroy the truth of the surfaces, that it may be lifted from its seat against the atmospheric pressure without the employment of 1 lb. per square inch more than that due to the weight of the valve; indeed, it is not found in practice to require anything more than the simple weight of the valve.

There can be but one explanation of this well-known practical fact, which is, that the valve does not fit sufficiently closely to its seat to exclude a thin film of air which acts to counterbalance the pressure on the back of the valve.

That such a film of air may exist without affecting the practical tightness of the valve is capable of indisputable demonstration. Thus, take two truly surfaced plates, and, after carefully cleaning them, lower one vertically upon the other, and it will be found to float upon the other, gliding about with extraordinary ease. Then lift the top plate, and it will suspend the bottom one from the partial vacuum between the two surfaces.

CONDITIONS UNDER WHICH A SLIDE VALVE OPERATES.

The conditions under which a slide valve operates while actually at work are widely different from those existing when it is at rest. First, as the boiler pressure fluctuates the temperature of the valve varies, causing the valve to expand or contract as the temperature increases or falls, and the irregular shape of the valve prevents the valve face from expanding or contracting in a straight line; hence the plane of its face is in a state of change constant with variations in the boiler pressure, and hence there is every probability, indeed, it appears a certainty, that there is at all times a film of steam beneath a part of the valve face acting to offset the pressure on the back of the valve. During an extensive experience in repairing locomotives, I never yet found an old valve a sufficiently good fit to its seat to have any of the atmospheric pressure holding it to its seat, and if air finds its way beneath why will not steam also?

Suppose that all slide valves were made of an equally good fit to their seats (and this is supposing a good deal, when we remember that some engine builders put in the valves just as they were planed, making no attempt to fit them to their seats on the cylinder port faces, while others file them to a fit, and others again scrape both valve and seat true to a surface plate). Suppose that the co-efficient of friction, whether due to the pressure only of the valve to its seat or to the combined pressure and induced adhesion from perfect contact, was in all cases alike, when the valves were put in new.

Let us see how long they would remain so. First, then, an iron or brass casting, heated after having the surface removed by planing or filing warps, and its fit is impaired. With the loss of the fit goes a loss of the adhesion, and an admission of steam beneath that part of the valve which does not fit. How much it will warp depends upon the temperature to which it is heated, on how much was cut off the planed face, on how unevenly the valve casting cooled after being taken out of the mould, on the shape and thickness of the valve, and on several other elements. Let us presume, however, that a casting could be made so that it would not warp from having its surface skin removed, and that, by heating the valve after it had been once surfaced, the reset had taken place, and the valve, being refaced true, would not again warp from being reheated (as experience demonstrates that it always does), and that, being heated to a given temperature, it would remain as close a fit to its seat as it was when cold. Then, just so soon as the temperature varied, the expansion and shape of the valve would vary. Cast iron expands by heat, in proportion to the temperature. The valve has, acting on the inside area of its exhaust port, the cooling effects of the atmosphere, which finds ingress through the exhaust pipe. The exhaust steam itself lowers in temperature as its pressure decreases, and the live steam on the back of the valve is comparatively constant in temperature : as a result, then, the valve is continually changing in form from the expansion due to the high temperature of the exhaust steam during the early part, and the lower temperature during the latter part, of the exhaust. Now comes another and more important question, and that is: How far will the spring of the valve, from the pressure of the steam upon its back, affect the fit to its seat, and will it so spring as to permit of a fine film of steam finding its way beneath the wings of the valve, thus relieving, to a certain extent, the amount of its pressure to its seat?

If we take a pair of the plates shown in Fig. 29, and get

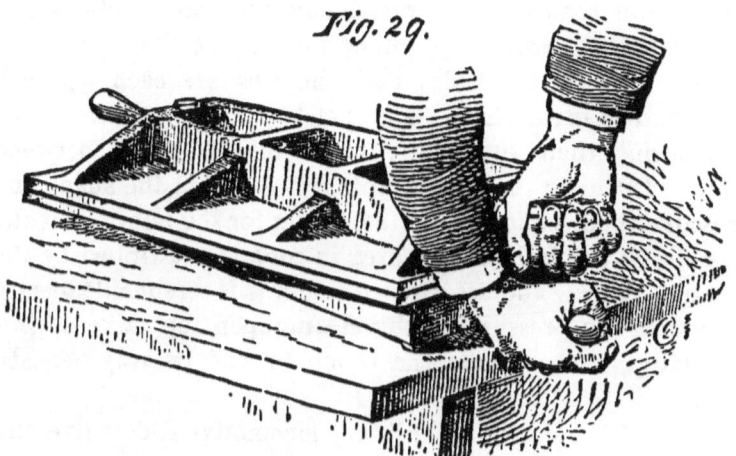

Fig. 29.

them so closely together that it requires, say, 340 lbs. to slide one upon the other, and then take hold of the plates by the handles, as shown in our engraving, we can pull them apart by exerting a force of about 130 lbs.; in other words, it will require but little more than one third as much power to pull them apart, in this manner, as it requires to slide one upon the other. In thus pulling them apart, we have, upon the back, whatever weight of the atmosphere the fineness of the fit leaves unbalanced, and, in addition, whatever amount of adhesion the perfect contact of the surfaces may induce. Hence, allowing a co-efficient of friction of 0·15, we should have 2,276 lbs. holding the plates to gether; and while allow ing a co-efficient of 23·7, we should have 1,440 lbs. resisting the effort to pull the plates apart. The fact, therefore, that 130 lbs. will actually, under the conditions shown, pull the plates apart, appears at first sight not a little singular. The solution, however, is simple enough. The plates spring from the pressure placed by the hands upon them, and hence they unlap and come apart just as if we took two sheets of paper, placed together and soaked with water, and then took hold of two corresponding corners and pulled them apart. The plates are $\frac{1}{2}$ inch thick in the body, and the ribs are each $\frac{7}{16}$ inch thick and $2\frac{3}{8}$ inches high; and yet 130 lbs. applied as shown will spring them sufficiently to let the air get in between them. Let us in the light of this fact examine the shape and pressure upon a slide valve (assuming for the nonce that the pressure is the unbalanced area in contact multiplied by the steam pressure), and ascertain whether it is reasonable to suppose that the pressure of the steam upon the valve spring the wings, and permits the steam to find its way beneath them.

In Fig. 30 is shown an ordinary locomotive slide valve, the

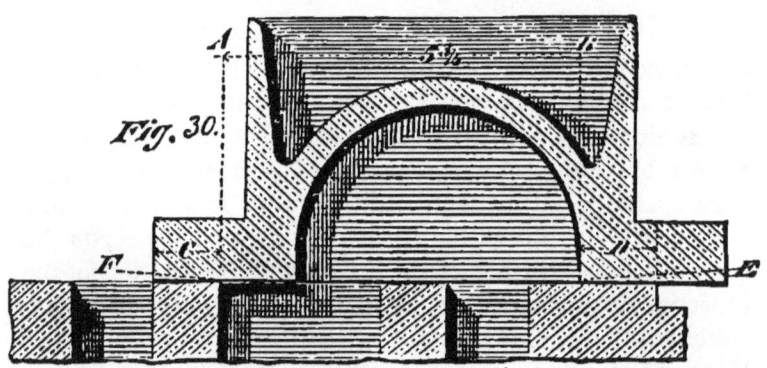

ports being $1\frac{1}{4}$ x 17 inches, the bridges between ports 1 inch wide, the cylinder exhaust port $2\frac{1}{2}$ inches wide, and the valve having 1 inch of steam lap, covering the ends of the cylinder ports 1 inch at each end. When the valve is in the position shown, it will be noted that there is a very large proportion of the area of the valve unsupported by the seat; the area of this portion will be in this case 5¾ inches, as marked in the engraving, one way, and 17 inches the other = 97·75 inches. Now supposing the steam pressure to be 130 lbs. per inch: then 97·75 × 130 = 12,707 lbs., the assumed pressure of the valve to its seat, tending to spring the flanges or wings in the direction denoted by the dotted lines, E and F, respectively. What have we to offset this amount? The area of one bridge equals 17, the area covered under the valve flange at D equals 11 inches, and the amount of the valve flange overlapping the ends of the steam ports equals 15·5; total 43·5 square inches, which, multiplied by the steam pressure, would give 5,855 lbs. as the pressure tending to keep the valve wings from springing. There will, it is true, be a pressure placed on the underneath side of the valve by the exhausting steam, the area thus acted on being, in the position shown, 97·79 square inches; but it can scarcely be advanced that this pressure can be sufficient to relieve the valve from its liability to spring from the 6,852 lbs. on the other side.

Theoretically, a valve will spring of its own weight; and that it will spring from the pressure which a man can put upon it with his hands, I have often found in facing valves up. For example, if, in trying the valve on the surface plate, the former is pressed in the middle by the hands to make the plate mark the face plainly, and the valve is fitted under these conditions to a practically perfect fit, the surface plate marks showing equally all over, we may then let the valve lie upon the plate of its own weight only, and the marks will show (after of course moving the valve back and forth) at and near the edges of the valve only, showing that the pressure of the hands sprung it. There are plenty of instances of metal in the most solid of forms springing of its own weight: witness the Morton Poole rolls, which, though of chilled cast iron and 12 inches in diameter, spring and bend by the insertion between them of a piece of gold leaf $\frac{1}{5000}$ inch thick. There is yet another part of this question, however, which is found in practice to be of the utmost importance, and that is (as a visit to any locomotive repair shop will demonstrate, by the engines that come in to be repaired), that the valve wears out of truth, and so does the seat. In my experience, I have chipped a full $\frac{1}{16}$ inch off valve seat faces without cutting the worn grooves out. I have examined, or had come under my observation, at least 400 slide valves, and I never saw one that was, after working three months, of a sufficient fit to its seat to require one pound more than its own weight to lift it from its seat; whereas, if such a valve as is shown in Fig. 30 were of a practically perfect fit, it would require, when in mid-position, some 800 lbs. to lift it vertically, taking hold of the ribs outside the arch. The fact is that the bridges wear hollow lengthways, and hollow, as denoted by a straight edge, over the seat and across the bridges. Then there usually wears in the seat face a groove at right angles to, and close to, the edges of the ports. To remedy this, a practice sprung up in England, in about the year 1865, of drilling, in the face of the valve and in a line with the exhaust-port edge, a hole in each wing; and this hole may be found mentioned in recent English engine specifications. Now just so soon as a valve face loses its smoothness, though the grooves may be only the one hundredth of an inch deep, or like coarse file marks, it becomes impracticable to exclude the surrounding air at atmospheric pressure, let alone steam at a high pressure, from between the surfaces.

I have a plate of the same size as those shown in Fig. 29, which has been planed and not fitted in any way. The planer marks are all intact. By placing a finished true plate upon it, the partial vacuum between the two will lift the planed one; but in about ten seconds it will fall, because the weight of the plate causes it to sag, and the air travels along the fine planer marks until there is not sufficient vacuum to sustain the weight of the plate, which is about 20 lbs. Now since the planed plate can be lifted by the vacuum, it is at least as good a fit as an ordinary slide valve, and under a steam pressure would undoubtedly be steam-tight, although the steam, like the air, would find its way along the planer marks, and thus counterbalance a large proportion of the pressure placed by the steam on the back of the valve. How much the elements of warping from expansion, changing form from irregular temperature, and counterbalancing from steam finding its way beneath the valve, will affect the pressure of a valve to its seat whether these causes act either in concert or partly counteract each the other, will depend upon the shape, size, strength, etc., of the valve.

COEFFICIENT OF FRICTION.

If we attempt to ascertain the power required to operate a slide valve by any of the rules given by various authors, we are at once confronted with several difficulties. First, we have no data as to how the friction of metal surfaces is affected by the metal being heated to various temperatures. Secondly, the dryness or wetness of the steam must have an important bearing, since steam acts to a certain extent as a lubricator, and the degree of that extent varies according to its saturation or wetness as it may be termed. Thirdly, we have no data which enables us to determine whether the closeness of the fit of the valve to its seat in those parts of its area which have no interposed film of steam beneath induces any adhesion of the surfaces which some experimenters claim exists in addition to the friction due to the pressure forcing the surfaces together. Since, however, this part of the subject will be referred to presently, we may now place before the reader the generally accepted table of coefficients of friction.

The experiments of General Morin on the friction of various bodies without an interposed film of lubricating liquid, but with the surfaces wiped clean by a greasy cloth, have been summarized by Professor Rankine in the following table.

	Angle of repose.	Friction in terms of the weight.
Metals on metals, dry	8½° to 11½°	0·15 to 0·2
Metals on metals, wet	16½°	0·3
Smooth surfaces, greased	4° to 4½°	0·07 to 0·08
Smooth surfaces, best results	1¾° to 2°	0·03 to 0·36

In a paper, of which an abstract has appeared in the *Comptes Rendus* of the French Academy of Sciences, for April 26, 1858, M. H. Bochet describes a series of experiments which have led him to the conclusion that the friction between a pair of surfaces of iron is not, as it has hitherto been believed, absolutely independent of the velocity of sliding, but that it diminishes slowly as that velocity increases.

If we class the conditions under which a slide valve operates under the head of "metals on metals, dry," we are confronted at once with the question: For what reason shall we select the co-efficient as 0·15 in preference to the 0·2, or *vice versa?* If we class those conditions under any other of the headings in the table, where are we to get a co-efficient of 0·15? And if, as M. H. Bochet concludes, the co-efficient varies with the velocity of sliding, how can we assume a fixed co-efficient for a slide valve when its velocity of sliding varies with every variation in the speed of the engine, as well as at every inch of its movement? In the case of slide valves, however, the weight upon the valve is not a dead weight, but live steam which will find its way beneath the surfaces and remain there.

In Appleton's "Cyclopædia" occurs the following: "Two glass or metal plates with well ground surfaces, when pressed together, will adhere with such force that the upper one will not only support the lower, but an additional weight will be required to separate them. The amount of this adhesive force has been measured by recording the weights necessary for their separation. The records of the old experimenters on this subject are worthless, because they placed a lubricating fluid (oil or fat) between the plates; they found thus the cohesion of the oil or fat, and not the adhesion of the plates. In later times, Prechtl, in Germany, has made the most careful experiments in this line ; he took polished metal plates of $1\frac{1}{2}$ inch diameter, suspended the upper one to a balance, brought it to an equilibrium in a horizontal position, and attached the lower plate to a support beneath it. Both plates were then brought into contact, so that the flat polished surfaces covered one another perfectly, and the weights required in the scale, at the other end of the balance beam, to separate the plates were the measures of adhesion. He found thus the following remarkable law : The adhesion between two plates of the same material is the same as that between one of the plates and any material which possesses a less adhesive force. Prechtl found also that an attraction of the plates manifested itself at an appreciable distance before actual contact, and he even measured the amount of this attraction at the distance of $\frac{1}{24}$ of an inch by means of weights in fractions of grains. The suspended plate, when brought within this distance, was attracted with an accelerated motion till the contact took place with a slight concussion. The idea that the pressure of the air was the chief cause of the adhesion of two such plates, as it is in the case of the well known experiment with the Madgeburg hemispheres, was set at rest by Boyle, who suspended the adhesive plates charged with weight in the vacuum of and air pump ; the plates were not separated, while the hemispheres held together by the vacuum alone fell apart."

In cast iron the amount of oil that will remain in the pores of the metal, even after careful wiping of the surface with dry rags, is sufficient to affect the friction of the surfaces, so that after such a surface has been oiled it is necessary to use alcohol in order to thoroughly remove the oil.

A practical experiment made by me upon a pair of cast-iron surface plates weighing 23 lbs. each and having an area of 96 inches demonstrated the following: With the top plate lowered vertically upon the bottom, it did not require an ounce to move the top plate, both plates being dry, but if there was one drop or one-quarter of a drop of oil distributed over the 192 inches of *area of the two plates*, it would require 50 to 100 lbs. to slide it, according to the amount of lubrication. This will always be the effect of slight lubrication upon closely fitting *smooth* surfaces, since the oil acted to exclude the film of air that would otherwise remain between the faces. Freely lubricated, the plate would require about 5 lbs. to slide it. If, however, the surfaces are wiped clean with oil and rag and are then wiped dry with the rag, and if we then place the plates in contact at one corner only, and slide the top one over the other, it has taken 341½ lbs. to slide one over the other, as appears from a test made by Messrs. Fairbanks upon their standard scales at the Centennial.

Now the area of the plate 96 inches multiplied by the atmospheric pressure per square inch $14·7 = 1411·2$, add the weight of the plate 23 lbs. and we have a total of 1434·2 lbs. pressing the plates together. Now since the power required to slide the plate expressed in hundreds is 3·415, we divide that amount into the total 1434·6, and obtain 23 as the power required to slide the top plate for every 100 lbs. pressing it to the bottom one, or, in other words, we obtain a coefficient of friction of ·23. This is higher than that given by Morin for cast iron or cast iron dry (as the plates were), and this could only arise from the surfaces being in more perfect contact or from the adhesion referred to. If in an experiment to determine the coefficient of friction the simple weight of the sliding piece is taken, it becomes an open question whether the surfaces exclude the air sufficiently to cause the atmospheric pressure to exert any influence whatever upon the external area, and if, on the other hand, the atmospheric pressure is taken in addition to the weight of the piece, it becomes a question how closely the surfaces fit together.

The size of the valve is a very important element, since it is obvious that in the case of a very large valve, such as is sometimes used in marine engine practice, it would be impracticable to make it strong enough to resist local distortion, causing it when under pressure to warp and fit more closely to its seat in those parts of the surface where the valve face, when not under pressure, would not fit to the seat.

It would take an enormous pressure to warp a block of cast iron 3 inches square and 2 inches thick sufficiently to destroy its close fit when the pressure was exerted all over its exposed area, even though it were supported, as on the bridges of a steam chest face, but were the piece 18 inches square a moderate pressure would make a sensible alteration; while a variation of 100° of temperature would affect even the small block to a sensible degree.

THE LUBRICATION OF SLIDE VALVES.

In many cases the lubrication of a slide valve, though necessary to prevent the surface from rapid abrasion, may act to increase the power necessary to operate the valve, unless, indeed, it be both copious and constant in supply. It has been stated that a fine film of oil acted to exclude the film of air from between the surfaces that were made by me as experiments on the surface plates before referred to, increasing the power required to move the top plate from 1 oz. up to 100 lbs., and this I have found in many experiments *always to be the case, the effect increasing with the smoothness and truth of the surfaces.* Furthermore, on a very smooth surface, and at the same time an *imporous* surface, the oil is rapidly wiped off by the motion, whereas, on one that appears bright and smooth to the eye but pitted when put under a magnifying glass of ordinary power, the effects of slight lubrication remain longer, probably because the oil remains longer in the minute depressions.

Now, with the continuous change of form shown in previous remarks to accompany the operation of a slide valve, the surfaces cannot remain in close contact, and the oil remains to some extent between those parts of the surfaces least in contact, hence lubrication undoubtedly assists the easy operation of the valve.

PRACTICAL METHOD OF FINDING THE WORKING RESULTS OF A GIVEN SLIDE VALVE.

A simple method of ascertaining, with a pair of compasses and a square, the working results of any given slide valve is shown in Fig. 31.

Suppose the given proportions of the valve are as follows:—

	Inches.
Width of steam ports	$\frac{5}{8}$
Width of cylinder exhaust port	$1\frac{1}{4}$
Steam lap on valve	$\frac{5}{8}$
Travel of valve	$2\frac{5}{8}$
Stroke of piston	24

Draw the circle A, its circumference representing the path of the crank-pin and its diameter representing the piston stroke, but since to draw it full size would be inconveniently large, we may draw it to a scale of $\frac{1}{8}$ inch per foot. Draw the circle B, whose diameter must equal the full stroke of the valve (in this case $2\frac{5}{8}$ inches), both circles being struck from the same centre. Now draw the vertical line C, which must pass through the centre of the circles. To the right of C, and at a distance from it equal to the lap of the valve, draw the line D.

Draw the small circle 1, which represents the crank-pin (on its dead-centre furthest from the engine-cylinder) ready to begin its stroke. Place one point of a pair of compasses on the centre of circle 1, and the other point (at the intersection

Fig. 31.

of D with the circle B) on B at E. Now, since one point of these compasses will always be applied to A, and the other to B, we may term them respectively the A and the B points.

The line D now represents two things: first the edge of the valve, the eccentric being set forward or advanced on the shaft sufficiently to take up the lap of the valve S, representing that angular advance, as it is termed (its amount being its angle from the line C).

We have now marked the relative positions of the crank-pin (at 1) on its dead centre and the position of the edge of the valve (at E), and, since the eccentric is fast to the crank-shaft, it is evident that their distances apart will remain the same at every point in the crank pin revolution.

Now mark the line F, whose length must equal, from the line D to its point of junction with the circle B, the width of the steam port, in this case $\frac{3}{8}$ inch, F being at a right angle to D.

Now while the eccentric moves in the arc of a circle from E to G, the edge of the valve moves in a line from D to G, and as F is the width of the port, the latter will be full open when it arrives at G. If, then, we place the B point of the compasses at G, and with the other point mark a point on the circle A, this point will denote the position of the crank-pin when the steam port (which is denoted in the figure by the inclosed square containing diagonal lines) is full open, as shown on the diagram at 2.

The diameter of the circle B being equal to the full stroke of valve, it is evident that from the point 4 we may mark off with the compasses the position of the crank-pin when the valve is at the end of its travel, such position being shown at 3.

The edge of the valve now stands at H but parallel to D, and it is obvious that when it retreats to I it will begin to close the latter, hence from I we mark A the position of the crank-pin, where the steam port begins to close.

When the edge of the valve meets the edge D of the steam port again, the latter will be closed, and the expansion of the steam in the cylinder will begin; hence we place the B point of compasses at J, and mark on A the position of crank-pin when the expansion begins, that position being shown at 5. The expansion will continue while the valve travels with its lap covering the steam port, which, being in this case ⅜ of an inch, is from J to K.

As the edge of the valve, which is parallel to the line C, at the point K leaves K, the port (which is still the same port F) opens for the exhaust; hence we mark the line L, whose length equals the width of the steam port, and which now represents the port as an exhaust port exactly as F did for a steam port.

From K we mark the crank-pin position when the exhaust begins, which is shown at 6.

When the valve-edge has moved across the port, the latter will be open full as an exhaust port; hence with the B point at M we mark on A the position 7 of crank-pin, it then being on its other dead-centre.

The exhaust is already full open, and it continues so while the valve passes to the end of its travel and returns to the edge of the port; hence we place the B point at N (the end of valve travel), and mark position of crank-pin at 8.

Continuing around the circle, we place compass-point at O, which represents the edge of the port denoted by L (which is virtually the same point in the valve movement as is the valve M), and mark on A position 9, which is the position of the crank-pin when the exhaust begins to close. With the compasses on B at P we mark on A point 10, which is the position of the crank-pin when the exhaust closes and the compression or cushioning of the steam in the cylinder begins, it continuing until the crank-pin arrives at position 1, from whence it started.

We have now to ascertain the position of the engine piston when the crank-pin was at each respective point, which is easily done, as follows:—

The full stroke of the piston is represented by the diameter of circle A, and hence by the distance between the vertical lines $Q R$, Q representing the end from which the piston started; hence we draw the vertical lines shown from each crank-pin position, and their distance from Q is the distance the piston has traveled when the crank-pin was in the respective positions; but as the circle A is one-eighth full size, we multiply the distance actually shown on the diagram by 8, and thus obtain from our diagram the following reading:—

Steam-port full open when piston had moved $3\frac{3}{8}$ inches.

Valve at end of travel when piston had moved $6\frac{3}{8}$ inches.

Steam-port began to close when piston had moved $9\frac{3}{8}$ inches.

Expansion began when piston had moved $18\frac{1}{2}$ inches.

Expansion ended and exhaust began when piston had moved $22\frac{1}{4}$ inches.

Port full open as an exhaust-port when piston arrived at end of stroke.

Exhaust remained open till piston had traveled $18\frac{1}{2}$ inches of return-stroke, when it began to close.

Exhaust closed when piston had moved $22\frac{1}{2}$ inches of return-stroke.

The compression and cushion took place during the last $1\frac{1}{2}$ inches of piston movement.

It is to be observed, however, that the angularity of the piston rod would cause a slight variation in the above data, the amount depending on the length of the connecting-rod in proportion to the length of the piston stroke. The effect of the variation would be to place positions 2, 3, and 4 further in advance than in the diagram, and positions 5 and 6 a little further back when the valve was in the respective positions;

while positions 9 and 10 would be placed slightly further to the left or in advance.

It will be noted that in this case we have not given the valve any lead; but all that is necessary to take lead into account is to mark to the right of D another line distant from D to the amount of the lead, in which case D will represent the edge of the port and the new line the edge of the valve. The compass points being set from 1 to the intersection of the new line with the circle A, this will throw the points 2, 3, 4, 5, and 6 further back towards 1. For the exhaust side we mark a new line to the left of line M, O, using its junction with B as the point of full port opening for exhaust; this will throw the point M nearer to 1 and the point O also nearer to 1. The positions of the port, as represented both at F and L, however, must remain with their nearest edges distant respectively from C to an amount equal to the lap only of the valve.

In our example no exhaust lap was given to the valve, but had it been otherwise we should require to mark to the left of line C another vertical line distant from C to the amount of exhaust lap, and the junction of this new line with the circle B would be the point wherefrom to mark off position 6, or, in other words, the position of the crank-pin when the exhaust-port opened, and also the point (in place of P) wherefrom to mark off crank-pin, position 10 where the compression or cushioning began, the position of port L remaining as in the diagram.

If, as is sometimes the case, a diagram thus laid out would have the circle B larger than A, if the latter be drawn to a scale of $\frac{1}{8}$, it may be drawn to a scale of $\frac{1}{4}$ or any other scale that will make A larger than B, but in any case the scale must be used as a multiplier in finding the position of the piston from the end of the stroke.

SETTING ECCENTRICS.

To set eccentrics upon a shaft before it is placed in the engine or to set them without moving the engine, proceed as follows: If the engine runs one way only, and therefore has but one eccentric, set the crank-shaft in such a position that a wooden straight-edge can be placed sufficiently beneath it to be out of the way of the eccentric. Place the centre line of the length of the crank truly horizontal, which may be done as follows: From the centre of the crank-shaft strike a circle of the diameter of the crank-pin, as shown in Fig. 32, at A, and draw upon the face of the crank a line that shall just meet the two circles as denoted by the line B, in Fig. 32, using a straight-edge, one end of which rests upon the crank-pin, while the other end is coincident with the perimeter of the circle A.

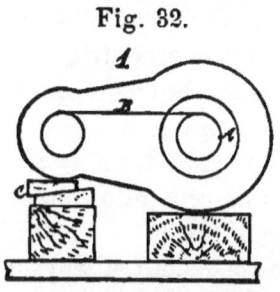

Fig. 32.

By means of the wedges shown at C, adjust the crank until the line B stands horizontally level, tested by a spirit-level. Then take a straight-edge, as in Fig. 33, and draw on it the line F, and place it horizontally level beneath the crank-shaft and the eccentric, fixing it temporarily so that it will not be liable to move. To find the centre of the crank-shaft G, hang over it plumb-lines as denoted by the lines $H I$, and where the plumb-lines intersect the line F, mark the points $J K$, and midway between $J K$ is the

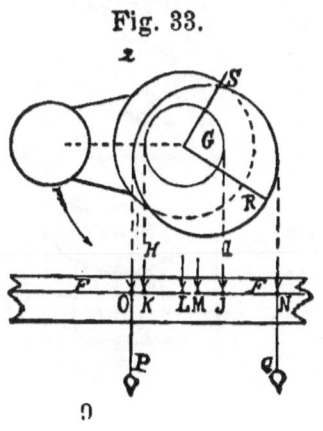

Fig. 33.

centre of the shaft marked on the straight-edge at L. From L, mark on the line F the point M, distant from L the amount equal to the lap of the valve added to the amount of lead the valve is required to have. From the point M mark on the line F the diameter of the eccentric producing the points N O; then hang over the circumference of the eccentric the plumb-lines P Q, and when these lines coincide with the points N O on F the eccentric is in its proper position.

Here, however, it is necessary to point out that in marking the point M it is necessary to consider which way the engine is to run. In Fig. 33 the arrow denotes the direction of the crank revolution, hence M is located on the right side of L, for M must, in an engine in which the slide valve gear has no rock-shaft, always be placed on the right of L farthest from the crank-pin, no matter at which end of the stroke the crank stands, or in which direction it is to run. But in engines having a rock-shaft, M must be placed on the side of L nearest to the crank-pin.

In Fig. 34 is shown the operation applied to an engine, such as a locomotive, having two eccentrics in order to enable it to run in either direction. In this case it is necessary to remember that the eccentric that is to operate the valve for the engine to run forward (that is in the direction denoted by the arrow in Fig. 34) must be the one that stands with its throw line following the crank, as shown at R in Fig. 34, in which the engine is supposed to have a rock-shaft, hence M is on the crank-pin side of L. The position of the backward eccentric, with relation to the crank, is precisely the same as eccentric R, save that its throw line, S, is as much one side of

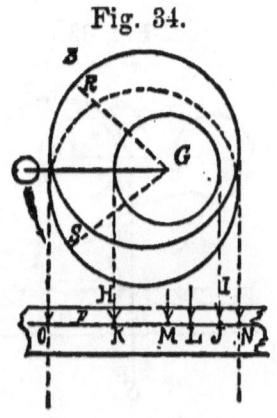

Fig. 34.

the crank as R's throw line is on the other, hence the plumblines P, Q, will intersect the points N, O, when placed on either eccentric. The eccentric that has its throw line in advance of the crank is called the leading one, hence in Fig. 34, S leads. In English engines the left hand eccentric leads, while in the United States the right hand one leads. If the wrong one is made to lead, the result will be that the engine will run backwards when the reversing lever is placed to run forward, and *vice versa*.

Referring again to Fig. 33, to cause the crank to revolve the other way all that would be necessary is to move the eccentric so that its throw line R stands in the direction denoted by S, hence R in Fig. 33 and R in Fig. 34 represent the eccentric set to run forward in both cases, one engine having, and the other not having, a rock-shaft. When the line of direction of the eccentric rod is not a line parallel with the centre line of the bore of the cylinder (which it usually is), the crank must be placed on the dead centre as before, but instead of the line F being adjusted to the centre-line of the crank, it must be adjusted to the centre-line of the eccentric-rod connection, the process being shown in Fig. 35, in which the engine is supposed to have a rock-shaft A connected to the slide spindle at one end, and to the eccentric-rod at the other. Hence the line of eccentric-rod connection is represented by the line B. Instead of carrying the centre-line of the shaft down vertically to the line F, standing horizontally, we must place the straight-edge at a right angle to B, and use a square as shown in the figure, instead of the plumb-lines, the rest of the operation remaining in all other respects the same.

Fig. 35.

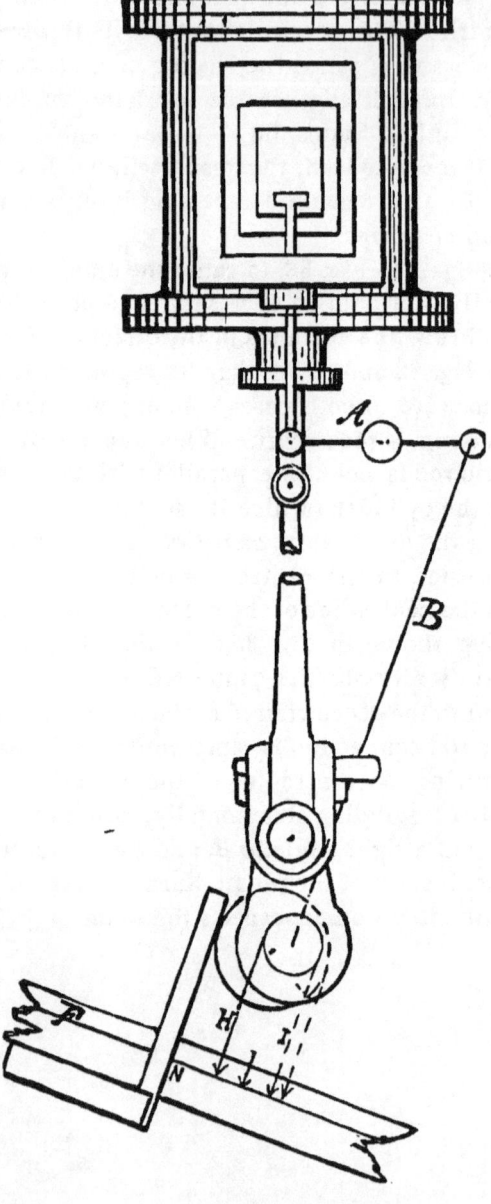

INDEX.

	PAGE
Angular advance or lead of an eccentric	9, 42, 43
Angularity of the connecting rod	18
Area of steam ports	63
Auchincloss, W. S.	70
Automatic cut-off slide valve	5
Bochet, M. H.	85
Bridges, width of	36, 68, 74
Bourne, John	68
Burgh, N. P.	69
Clearance, action of	31
Clearance, description of	30
Coefficient of friction	84–88
Compression	59, 63
Conditions under which a slide-valve operates	77
Connecting rod, angularity of	18, 43
Connecting rod, effects of angularity of	20
Crank movement	21
Crank, quarter revolutions of	21
Crank, speed of	19
Cut-off, point of	37
Cut-off slide valve	5
Cylinder, exhaust, contraction of	27, 35
Eccentric, angular advance of	8, 42, 43
Eccentric, effect of valve travel on the	34
Eccentric, piston and crank movements	21
Eccentrics, setting double	96
Eccentrics, setting single	95
Expansion, advantages of	14
Exhaust area during different periods of the stroke	65
Exhaust area in proportion to steam area	63, 66
Exhaust, back pressure on	54
Exhaust lap, when employed	29

	PAGE
Exhaust side of the valve, lap on	25
Expansion, comparisons of	60
Expansion of steam in the steam passages	61
Friction, coefficient of	84–88
Lap, action of steam	15
Lap, advantages of steam	15
Lap, exhaust	9, 25, 26, 27
Lap, exhaust, when employed	29
Lap, measurement of	14
Lap of a valve	13
Lap of the steam side of the valve	13
Lap, steam	13, 25
Lap, steam, the effects of	26, 27
Lap, without	5
Lead of an eccentric	9, 42, 43
Lead of valve, determination of	8
Locomotive practice, examples of valve movements from	44–56
Locomotive slide valve, illustration of	80
Locomotives, use of slide valve in	5
Lubrication of slide valves	83, 89
Piston, eccentric and crank movements	21
Piston, variations in velocity of	19, 22
Point of cut-off	37
Point of release	38
Ports, areas, proportion of exhaust to steam	66
Ports, effective area of	28
Ports, exhaust	6
Ports, opening of	24
Ports, proportioning area of	59, 68, 72
Ports, shape of	68
Ports, steam	6
Ports, variations in opening and closure of	17
Power required to operate a slide valve	76

(99)

	PAGE
Results of comparison of valve movements	57–68
Rock-shaft, diagram showing effects of	41
Rock-shaft, effect of, on eccentric position	40
Setting eccentrics	95–98
Shape of steam ports	68
Slide valve, engines to which applied	5
Slide valve, meaning of term	5
Slide valve, simplest form of	5
Stationary engines, automatic cut-off valves in	5
Steam, expansion of	15, 23
Steam lap	13
Steam ports, area and shape of	68
Steam ports, proportioning	72
Steam ports, shape of	68
Steam side of the valve, lap of	13
Steam supply	9
Steam, using expansively	14
Tab'es of comparison of valve movements	58, 59, 60, 61, 62, 63, 65, 66, 68
Valve, conditions under which it operates	77
Valve, diagram of action of	10
Valve, illustration of an ordinary locomotive	80
Valve, lap on the exhaust side of	26
Valve, lead	7, 8, 9
Valve movements, comparisons of	57, 58, 59, 60, 61, 62, 63, 64, 65, 66, 67, 68
Valve movements, examples of	44, 45, 46, 47, 48, 49, 50, 51, 52, 53, 54, 55, 56
Valve travel	7, 32, 51, 75, 29
Valve, wearing surface of a	67
Valves, comparisons of different	39
Valves, finding working results of	89–94
Valves, fit of, to their seats	77
Valves, friction of	84
Valves, lubrication of	88, 89
Valves, power required to move	73, 76
Valves, pressure on	80
Valves, spring of	81
Valves, surfaces of	77
Valves, warping of	78, 79
Width of bridges	36, 68, 74
Without lap	5
Working results of a slide valve, method of finding	89–94
Zenner, Dr.	69

CATALOGUE
OF
PRACTICAL AND SCIENTIFIC BOOKS,

PUBLISHED BY

HENRY CAREY BAIRD & CO.,

Industrial Publishers and Booksellers,

NO. 810 WALNUT STREET,

PHILADELPHIA.

☞ Any of the Books comprised in this Catalogue will be sent by mail, free of postage, at the publication price.

☞ A Descriptive Catalogue, 96 pages, 8vo., will be sent, free of postage, to any one who will furnish the publisher with his address.

ARLOT.—A Complete Guide for Coach Painters.
Translated from the French of M. ARLOT, Coach Painter; for eleven years Foreman of Painting to M. Eberler, Coach Maker, Paris. By A. A. FESQUET, Chemist and Engineer. To which is added an Appendix, containing Information respecting the Materia's and the Practice of Coach and Car Painting and Varnishing in the United States and Great Britain. 12mo. $1.25

ARMENGAUD, AMOROUX, and JOHNSON.—The Practical Draughtsman's Book of Industrial Design, and Machinist's and Engineer's Drawing Companion:
Forming a Complete Course of Mechanical Engineering and Architectural Drawing. From the French of M. Armengaud the elder, Prof. of Design in the Conservatoire of Arts and Industry, Paris, and MM. Armengaud the younger, and Amoroux, Civil Engineers. Rewritten and arranged with additional matter and plates, selections from and examples of the most useful and generally employed mechanism of the day. By WILLIAM JOHNSON, Assoc. Inst. C. E., Editor of "The Practical Mechanic's Journal." Illustrated by 50 folio steel plates, and 50 wood-cuts. A new edition, 4to. $10.00

ARROWSMITH.—Paper-Hanger's Companion:
A Treatise in which the Practical Operations of the Trade are Systematically laid down: with Copious Directions Preparatory to Papering; Preventives against the Effect of Damp on Walls; the Various Cements and Pastes Adapted to the Several Purposes of the Trade; Observations and Directions for the Panelling and Ornamenting of Rooms, etc. By JAMES ARROWSMITH, Author of "Analysis of Drapery," etc. 12mo., cloth. $1.25

ASHTON.—The Theory and Practice of the Art of Designing Fancy Cotton and Woollen Cloths from Sample:
Giving full Instructions for Reducing Drafts as well as the Methods of Spooling and Making out Harness for Cross Drafts, and Finding any Required Reed, with Calculations and Tables of Yarn. By FREDERICK T. ASHTON, Designer, West Pittsfield, Mass. With 52 Illustrations. One volume, 4to. $10.00

BAIRD.—Letters on the Crisis, the Currency and the Credit System.
By HENRY CAREY BAIRD. Pamphlet. 05

BAIRD.—Protection of Home Labor and Home Productions necessary to the Prosperity of the American Farmer.
By HENRY CAREY BAIRD. 8vo., paper. 10

BAIRD.—Some of the Fallacies of British Free-Trade Revenue Reform.
Two Letters to Arthur Latham Perry, Professor of History and Political Economy in Williams College. By HENRY CAREY BAIRD. Pamphlet. 05

BAIRD.—The Rights of American Producers, and the Wrongs of British Free-Trade Revenue Reform.
By HENRY CAREY BAIRD. Pamphlet. 05

BAIRD.—Standard Wages Computing Tables:
An Improvement in all former Methods of Computation, so arranged that wages for days, hours, or fractions of hours, at a specified rate per day or hour, may be ascertained at a glance. By T. SPANGLER BAIRD. Oblong folio. $5.00

BAIRD.—The American Cotton Spinner, and Manager's and Carder's Guide:
A Practical Treatise on Cotton Spinning; giving the Dimensions and Speed of Machinery, Draught and Twist Calculations, etc.; with notices of recent Improvements: together with Rules and Examples for making changes in the sizes and numbers of Roving and Yarn. Compiled from the papers of the late ROBERT H. BAIRD. 12mo. $1.50

BAKER.—Long-Span Railway Bridges :
Comprising Investigations of the Comparative Theoretical and Practical Advantages of the various Adopted or Proposed Type Systems of Construction; with numerous Formulæ and Tables. By B. BAKER. 12mo. $2.00

BAUERMAN.—A Treatise on the Metallurgy of Iron :
Containing Outlines of the History of Iron Manufacture, Methods of Assay, and Analysis of Iron Ores, Processes of Manufacture of Iron and Steel, etc., etc. By H. BAUERMAN, F. G. S., Associate of the Royal School of Mines. First American Edition, Revised and Enlarged. With an Appendix on the Martin Process for Making Steel, from the Report of ABRAM S. HEWITT, U. S. Commissioner to the Universal Exposition at Paris, 1867. Illustrated. 12mo. . $2.00

BEANS.—A Treatise on Railway Curves and the Location of Railways.
By E. W. BEANS, C. E. Illustrated. 12mo. Tucks. . . $1.50

BELL.—Carpentry Made Easy :
Or, The Science and Art of Framing on a New and Improved System. With Specific Instructions for Building Balloon Frames, Barn Frames, Mill Frames, Warehouses, Church Spires, etc. Comprising also a System of Bridge Building, with Bills, Estimates of Cost, and valuable Tables. Illustrated by 38 plates, comprising nearly 200 figures. By WILLIAM E. BELL, Architect and Practical Builder. 8vo. . $5.00

BELL.—Chemical Phenomena of Iron Smelting :
An Experimental and Practical Examination of the Circumstances which determine the Capacity of the Blast Furnace, the Temperature of the Air, and the proper Condition of the Materials to be operated upon. By I. LOWTHIAN BELL. Illustrated. 8vo. . . $6.00

BEMROSE.—Manual of Wood Carving :
With Practical Illustrations for Learners of the Art, and Original and Selected Designs. By WILLIAM BEMROSE, Jr. With an Introduction by LLEWELLYN JEWITT, F. S. A., etc. With 128 Illustrations. 4to., cloth. $3.00

BICKNELL.—Village Builder, and Supplement :
Elevations and Plans for Cottages, Villas, Suburban Residences, Farm Houses, Stables and Carriage Houses. Store Fronts, School Houses, Churches, Court Houses, and a model Jail; also, Exterior and Interior details for Public and Private Buildings, with approved Forms of Contracts and Specifications, including Prices of Building Materials and Labor at Boston, Mass., and St. Louis, Mo. Containing 75 plates drawn to scale; showing the style and cost of building in different sections of the country, being an original work comprising the designs of twenty leading architects, representing the New England, Middle, Western, and Southwestern States. 4to. . $10.00

BLENKARN.—Practical Specifications of Works executed in Architecture, Civil and Mechanical Engineering, and in Road Making and Sewering:
To which are added a series of practically useful Agreements and Reports. By JOHN BLENKARN. Illustrated by 15 large folding plates. 8vo. $9.00

BLINN.—A Practical Workshop Companion for Tin, Sheet-Iron, and Copperplate Workers:
Containing Rules for describing various kinds of Patterns used by Tin, Sheet-Iron, and Copper-plate Workers; Practical Geometry; Mensuration of Surfaces and Solids; Tables of the Weights of Metals, Lead Pipe, etc.; Tables of Areas and Circumferences of Circles; Japan, Varnishes, Lackers, Cements, Compositions, etc., etc. By LEROY J. BLINN, Master Mechanic. With over 100 Illustrations. 12mo. $2.50

BOOTH.—Marble Worker's Manual:
Containing Practical Information respecting Marbles in general, their Cutting, Working, and Polishing; Veneering of Marble; Mosaics; Composition and Use of Artificial Marble, Stuccos, Cements, Receipts, Secrets, etc., etc. Translated from the French by M. L. BOOTH. With an Appendix concerning American Marbles. 12mo., cloth. $1.50

BOOTH AND MORFIT.—The Encyclopedia of Chemistry, Practical and Theoretical:
Embracing its application to the Arts, Metallurgy, Mineralogy, Geology, Medicine, and Pharmacy. By JAMES C. BOOTH, Melter and Refiner in the United States Mint, Professor of Applied Chemistry in the Franklin Institute, etc., assisted by CAMPBELL MORFIT, author of "Chemical Manipulations," etc. Seventh edition. Royal 8vo., 978 pages, with numerous wood-cuts and other illustrations. . £5.00

BOX.—A Practical Treatise on Heat:
As applied to the Useful Arts; for the Use of Engineers, Architects, etc. By THOMAS BOX, author of "Practical Hydraulics." Illustrated by 14 plates containing 114 figures. 12mo. $5.00

BOX.—Practical Hydraulics:
A Series of Rules and Tables for the use of Engineers, etc. By THOMAS BOX. 12mo. $2.50

BROWN.—Five Hundred and Seven Mechanical Movements:
Embracing all those which are most important in Dynamics, Hydraulics, Hydrostatics, Pneumatics, Steam Engines, Mill and other Gearing, Presses, Horology, and Miscellaneous Machinery; and including many movements never before published, and several of which have only recently come into use. By HENRY T. BROWN, Editor of the "American Artisan." In one volume, 12mo. . . . $1.00

HENRY CAREY BAIRD'S CATALOGUE. 5

BUCKMASTER.—The Elements of Mechanical Physics:
By J. C. BUCKMASTER, late Student in the Government School of Mines; Certified Teacher of Science by the Department of Science and Art; Examiner in Chemistry and Physics in the Royal College of Preceptors; and late Lecturer in Chemistry and Physics of the Royal Polytechnic Institute. Illustrated with numerous engravings. In one volume, 12mo. $1.50

BULLOCK.—The American Cottage Builder:
A Series of Designs, Plans, and Specifications, from $200 to $20,000, for Homes for the People; together with Warming, Ventilation, Drainage, Painting, and Landscape Gardening. By JOHN BULLOCK, Architect, Civil Engineer, Mechanician, and Editor of "The Rudiments of Architecture and Building," etc., etc. Illustrated by 75 engravings. In one volume, 8vo. $3.50

BULLOCK.—The Rudiments of Architecture and Building:
For the use of Architects, Builders, Draughtsmen, Machinists, Engineers, and Mechanics. Edited by JOHN BULLOCK, author of "The American Cottage Builder." Illustrated by 250 engravings. In one volume, 8vo. $3.50

BURGH.—Practical Illustrations of Land and Marine Engines:
Showing in detail the Modern Improvements of High and Low Pressure, Surface Condensation, and Super-heating, together with Land and Marine Boilers. By N. P. BURGH, Engineer. Illustrated by 20 plates, double elephant folio, with text. $21.00

BURGH.—Practical Rules for the Proportions of Modern Engines and Boilers for Land and Marine Purposes.
By N. P. BURGH, Engineer. 12mo. $1.50

BURGH.—The Slide-Valve Practically Considered.
By N. P. BURGH, Engineer. Completely illustrated. 12mo. $2.00

BYLES.—Sophisms of Free Trade and Popular Political Economy Examined.
By a BARRISTER (Sir JOHN BARNARD BYLES, Judge of Common Pleas). First American from the Ninth English Edition, as published by the Manchester Reciprocity Association. In one volume, 12mo. Paper, 75 cts. Cloth $1.25

BYRN.—The Complete Practical Brewer:
Or Plain, Accurate, and Thorough Instructions in the Art of Brewing Beer, Ale, Porter, including the Process of making Bavarian Beer, all the Small Beers, such as Root-beer, Ginger-pop, Sarsaparilla-beer, Mead, Spruce Beer, etc., etc. Adapted to the use of Public Brewers and Private Families. By M. LA FAYETTE BYRN, M D. With illustrations. 12mo. $1.25

BYRN.—The Complete Practical Distiller:
Comprising the most perfect and exact Theoretical and Practical Description of the Art of Distillation and Rectification; including all of the most recent improvements in distilling apparatus; instructions for preparing spirits from the numerous vegetables, fruits, etc.; directions for the distillation and preparation of all kinds of brandies and other spirits, spirituous and other compounds, etc., etc. By M. LA FAYETTE BYRN, M. D. Eighth Edition. To which are added, Practical Directions for Distilling, from the French of Th. Fling, Brewer and Distiller. 12mo. $1.50

BYRNE.—Handbook for the Artisan, Mechanic, and Engineer:
Comprising the Grinding and Sharpening of Cutting Tools, Abrasive Processes, Lapidary Work, Gem and Glass Engraving, Varnishing and Lackering, Apparatus, Materials and Processes for Grinding and Polishing, etc. By OLIVER BYRNE. Illustrated by 185 wood engravings. In one volume, 8vo. $5.00

BYRNE.—Pocket Book for Railroad and Civil Engineers:
Containing New, Exact, and Concise Methods for Laying out Railroad Curves, Switches, Frog Angles, and Crossings; the Staking out of work; Levelling; the Calculation of Cuttings; Embankments; Earth-work, etc. By OLIVER BYRNE. 18mo., full bound, pocket-book form. $1.75

BYRNE.—The Practical Model Calculator:
For the Engineer, Mechanic, Manufacturer of Engine Work, Naval Architect, Miner, and Millwright. By OLIVER BYRNE. 1 volume, 8vo., nearly 600 pages $4.50

BYRNE.—The Practical Metal-Worker's Assistant:
Comprising Metallurgic Chemistry; the Arts of Working all Metals and Alloys; Forging of Iron and Steel; Hardening and Tempering; Melting and Mixing; Casting and Founding; Works in Sheet Metal; The Processes Dependent on the Ductility of the Metals; Soldering; and the most Improved Processes and Tools employed by Metal-Workers. With the Application of the Art of Electro-Metallurgy to Manufacturing Processes; collected from Original Sources, and from the Works of Holtzapffel, Bergeron, Leupold, Plumier, Napier, Scoffern, Clay, Fairbairn, and others. By OLIVER BYRNE. A new, revised, and improved edition, to which is added An Appendix, containing THE MANUFACTURE OF RUSSIAN SHEET-IRON. By JOHN PERCY, M. D., F.R.S. THE MANUFACTURE OF MALLEABLE IRON CASTINGS, and IMPROVEMENTS IN BESSEMER STEEL. By A. A. FESQUET, Chemist and Engineer. With over 600 Engravings, illustrating every Branch of the Subject. 8vo. $7.00

Cabinet Maker's Album of Furniture:
Comprising a Collection of Designs for Furniture. Illustrated by 48 Large and Beautifully Engraved Plates. In one vol., oblong $3.50

CALLINGHAM.—Sign Writing and Glass Embossing:
A Complete Practical Illustrated Manual of the Art. By JAMES CALLINGHAM. In one volume, 12mo. $1.50

CAMPIN.—A Practical Treatise on Mechanical Engineering:
Comprising Metallurgy, Moulding, Casting, Forging, Tools, Workshop Machinery, Mechanical Manipulation, Manufacture of Steam-engines, etc., etc. With an Appendix on the Analysis of Iron and Iron Ores. By FRANCIS CAMPIN, C. E. To which are added, Observations on the Construction of Steam Boilers, and Remarks upon Furnaces used for Smoke Prevention; with a Chapter on Explosions. By R. Armstrong, C. E., and John Bourne. Rules for Calculating the Change Wheels for Screws on a Turning Lathe, and for a Wheel-cutting Machine. By J. LA NICCA. Management of Steel, Including Forging, Hardening, Tempering, Annealing, Shrinking, and Expansion. And the Case-hardening of Iron. By G. EDE. 8vo. Illustrated with 29 plates and 100 wood engravings . . . $6.00

CAMPIN.—The Practice of Hand-Turning in Wood, Ivory, Shell, etc.:
With Instructions for Turning such works in Metal as may be required in the Practice of Turning Wood, Ivory, etc. Also, an Appendix on Ornamental Turning. By FRANCIS CAMPIN; with Numerous Illustrations. 12mo., cloth $3.00

CAREY.—The Works of Henry C. Carey:
FINANCIAL CRISES, their Causes and Effects. 8vo. paper . 25
HARMONY OF INTERESTS: Agricultural, Manufacturing, and Commercial. 8vo., cloth $1.50
MANUAL OF SOCIAL SCIENCE. Condensed from Carey's "Principles of Social Science." By KATE MCKEAN. 1 vol. 12mo. $2.25
MISCELLANEOUS WORKS: comprising "Harmony of Interests," "Money," "Letters to the President," "Financial Crises," "The Way to Outdo England Without Fighting Her," "Resources of the Union," "The Public Debt," "Contraction or Expansion?" "Review of the Decade 1857-'67," "Reconstruction," etc., etc. Two vols., 8vo., cloth . . . .
PAST, PRESENT, AND FUTURE. 8vo. $2.50
PRINCIPLES OF SOCIAL SCIENCE. 3 vols., 8vo., cloth $10.00
THE SLAVE-TRADE, DOMESTIC AND FOREIGN; Why it Exists, and How it may be Extinguished (1853). 8vo., cloth . $2.00
LETTERS ON INTERNATIONAL COPYRIGHT (1867) . 50
THE UNITY OF LAW: As Exhibited in the Relations of Physical, Social, Mental, and Moral Science (1872). In one volume, 8vo., pp. xxiii., 433. Cloth $3.50

CHAPMAN.—A Treatise on Ropemaking:
As Practised in private and public Rope yards, with a Description of the Manufacture, Rules, Tables of Weights, etc., adapted to the Trades, Shipping, Mining, Railways, Builders, etc. By ROBERT CHAPMAN. 24mo. $1.50

COLBURN.—The Locomotive Engine:
Including a Description of its Structure, Rules for Estimating its Capabilities, and Practical Observations on its Construction and Management. By ZERAH COLBURN. Illustrated. A new edition. 12mo. $1.25

CRAIK.—The Practical American Millwright and Miller.
By DAVID CRAIK, Millwright. Illustrated by numerous wood engravings, and two folding plates. 8vo. $5.00

DE GRAFF.—The Geometrical Stair Builders' Guide:
Being a Plain Practical System of Hand-Railing, embracing all its necessary Details, and Geometrically Illustrated by 22 Steel Engravings; together with the use of the most approved principles of Practical Geometry. By SIMON DE GRAFF, Architect. 4to. . $5.00

DE KONINCK.—DIETZ.—A Practical Manual of Chemical Analysis and Assaying:
As applied to the Manufacture of Iron from its Ores, and to Cast Iron, Wrought Iron, and Steel, as found in Commerce. By L. L. DE KONINCK, Dr. Sc., and E. DIETZ, Engineer. Edited with Notes, by ROBERT MALLET, F.R.S., F.S.G., M.I.C.E., etc. American Edition, Edited with Notes and an Appendix on Iron Ores, by A. A. FESQUET, Chemist and Engineer. One volume, 12mo. $2.50

DUNCAN.—Practical Surveyor's Guide:
Containing the necessary information to make any person, of common capacity, a finished land surveyor without the aid of a teacher. By ANDREW DUNCAN. Illustrated. 12mo., cloth. . . . $1.25

DUPLAIS.—A Treatise on the Manufacture and Distillation of Alcoholic Liquors:
Comprising Accurate and Complete Details in Regard to Alcohol from Wine, Molasses, Beets, Grain, Rice, Potatoes, Sorghum, Asphodel, Fruits, etc.; with the Distillation and Rectification of Brandy, Whiskey, Rum, Gin, Swiss Absinthe, etc., the Preparation of Aromatic Waters, Volatile Oils or Essences, Sugars, Syrups, Aromatic Tinctures, Liqueurs, Cordial Wines, Effervescing Wines, etc., the Aging of Brandy and the Improvement of Spirits, with Copious Directions and Tables for Testing and Reducing Spirituous Liquors, etc., etc. Translated and Edited from the French of MM. DUPLAIS, Ainé et Jeune. By M. McKENNIE, M.D. To which are added the United States Internal Revenue Regulations for the Assessment and Collection of Taxes on Distilled Spirits. Illustrated by fourteen folding plates and several wood engravings. 743 pp., 8vo. $10.00

DUSSAUCE.—A General Treatise on the Manufacture of Every Description of Soap:
Comprising the Chemistry of the Art, with Remarks on Alkalies, Saponifiable Fatty Bodies, the apparatus necessary in a Soap Factory, Practical Instructions in the manufacture of the various kinds of Soap, the assay of Soaps, etc., etc. Edited from Notes of Larmé, Fontenelle, Malapayre, Dufour, and others, with large and important additions by Prof. H. DUSSAUCE, Chemist. Illustrated. In one vol., 8vo. . $12.50

HENRY CAREY BAIRD'S CATALOGUE. 9

DUSSAUCE.—A General Treatise on the Manufacture of Vinegar:
Theoretical and Practical. Comprising the various Methods, by the Slow and the Quick Processes, with Alcohol, Wine, Grain, Malt, Cider, Molasses, and Beets; as well as the Fabrication of Wood Vinegar, etc., etc. By Prof. H. DUSSAUCE. In one volume, 8vo. . . $5.00

DUSSAUCE.—A New and Complete Treatise on the Arts of Tanning, Currying, and Leather Dressing:
Comprising all the Discoveries and Improvements made in France, Great Britain, and the United States. Edited from Notes and Documents of Messrs. Sallerou, Grouvelle, Duval, Dessables, Labarraque, Payen, René, De Fontenelle, Malapeyre, etc., etc. By Prof. H. DUSSAUCE, Chemist. Illustrated by 212 wood engravings. 8vo. $25.00

DUSSAUCE.—A Practical Guide for the Perfumer:
Being a New Treatise on Perfumery, the most favorable to the Beauty without being injurious to the Health, comprising a Description of the substances used in Perfumery, the Formulæ of more than 1000 Preparations, such as Cosmetics, Perfumed Oils, Tooth Powders, Waters, Extracts, Tinctures, Infusions, Spirits, Vinaigres, Essential Oils, Pastels, Creams, Soaps, and many new Hygienic Products not hitherto described. Edited from Notes and Documents of Messrs. Debay, Lunel, etc. With additions by Prof. H. DUSSAUCE, Chemist. 12mo.

DUSSAUCE.—Practical Treatise on the Fabrication of Matches, Gun Cotton, and Fulminating Powders.
By Prof. H. DUSSAUCE. 12mo. $3.00

Dyer and Color-maker's Companion:
Containing upwards of 200 Receipts for making Colors, on the most approved principles, for all the various styles and fabrics now in existence; with the Scouring Process, and plain Directions for Preparing, Washing-off, and Finishing the Goods. In one vol., 12mo. . $1.25

EASTON.—A Practical Treatise on Street or Horse-power Railways.
By ALEXANDER EASTON, C. E. Illustrated by 23 plates. 8vo., cloth. $3.00

ELDER.—Questions of the Day:
Economic and Social. By Dr. WILLIAM ELDER. 8vo. . $3.00

FAIRBAIRN.—The Principles of Mechanism and Machinery of Transmission:
Comprising the Principles of Mechanism, Wheels, and Pulleys, Strength and Proportions of Shafts, Coupling of Shafts, and Engaging and Disengaging Gear. By Sir WILLIAM FAIRBAIRN, C.E., LL.D., F.R.S., F.G.S. Beautifully illustrated by over 150 wood-cuts. In one volume, 12mo. $2.50

FORSYTH.—Book of Designs for Headstones, Mural, and other Monuments:
Containing 78 Designs. By JAMES FORSYTH. With an Introduction by CHARLES BOUTELL, M. A. 4to., cloth. $5.00

GIBSON.—The American Dyer:
A Practical Treatise on the Coloring of Wool, Cotton, Yarn and Cloth, in three parts. Part First gives a descriptive account of the Dye Stuffs; if of vegetable origin, where produced, how cultivated, and how prepared for use; if chemical, their composition, specific gravities, and general adaptability, how adulterated, and how to detect the adulterations, etc. Part Second is devoted to the Coloring of Wool, giving recipes for one hundred and twenty-nine different colors or shades, and is supplied with sixty colored samples of Wool. Part Third is devoted to the Coloring of Raw Cotton or Cotton Waste, for mixing with Wool Colors in the Manufacture of all kinds of Fabrics, gives recipes for thirty-eight different colors or shades, and is supplied with twenty-four colored samples of Cotton Waste. Also, recipes for Coloring Beavers, Doeskins, and Flannels, with remarks upon Anilines, giving recipes for fifteen different colors or shades, and nine samples of Aniline Colors that will stand both the Fulling and Scouring process. Also, recipes for Aniline Colors on Cotton Thread, and recipes for Common Colors on Cotton Yarns. Embracing in all over two hundred recipes for Colors and Shades, and ninety-four samples of Colored Wool and Cotton Waste, etc. By RICHARD H. GIBSON, Practical Dyer and Chemist. In one volume, 8vo. . . $6.00

GILBART.—History and Principles of Banking:
A Practical Treatise. By JAMES W. GILBART, late Manager of the London and Westminster Bank. With additions. In one volume, 8vo., 600 pages, sheep $5.00

Gothic Album for Cabinet Makers:
Comprising a Collection of Designs for Gothic Furniture. Illustrated by 23 large and beautifully engraved plates. Oblong . . $2.00

GRANT.—Beet-root Sugar and Cultivation of the Beet.
By E. B. GRANT. 12mo. $1.25

GREGORY.—Mathematics for Practical Men:
Adapted to the Pursuits of Surveyors, Architects, Mechanics, and Civil Engineers. By OLINTHUS GREGORY. 8vo., plates, cloth $3.00

GRISWOLD.—Railroad Engineer's Pocket Companion for the Field:
Comprising Rules for Calculating Deflection Distances and Angles, Tangential Distances and Angles, and all Necessary Tables for Engineers; also the art of Levelling from Preliminary Survey to the Construction of Railroads, intended Expressly for the Young Engineer, together with Numerous Valuable Rules and Examples. By W. GRISWOLD. 12mo., tucks $1.75

GRUNER.—Studies of Blast Furnace Phenomena.
By M. L. GRUNER, President of the General Council of Mines of France, and lately Professor of Metallurgy at the Ecole des Mines. Translated, with the Author's sanction, with an Appendix, by L. D. B. Gordon, F. R. S. E.. F. G. S. Illustrated. 8vo. . . . $2.50

HENRY CAREY BAIRD'S CATALOGUE. 11

GUETTIER.—Metallic Alloys:

Being a Practical Guide to their Chemical and Physical Properties, their Preparation, Composition, and Uses. Translated from the French of A. GUETTIER, Engineer and Director of Foundries, author of "La Fouderie en France," etc., etc. By A. A. FESQUET, Chemist and Engineer. In one volume, 12mo. $3.00

HARRIS.—Gas Superintendent's Pocket Companion.

By HARRIS & BROTHER, Gas Meter Manufacturers, 1115 and 1117 Cherry Street, Philadelphia. Full bound in pocket-book form $1.00

Hats and Felting:

A Practical Treatise on their Manufacture. By a Practical Hatter. Illustrated by Drawings of Machinery, etc. 8vo. . . . $1.25

HOFMANN.—A Practical Treatise on the Manufacture of Paper in all its Branches.

By CARL HOFMANN. Late Superintendent of paper mills in Germany and the United States; recently manager of the Public Ledger Paper Mills, near Elkton, Md. Illustrated by 110 wood engravings, and five large folding plates. In one volume, 4to., cloth; 398 pages $15.00

HUGHES.—American Miller and Millwright's Assistant.

By WM. CARTER HUGHES. A new edition. In one vol., 12mo. $1.50

HURST.—A Hand-Book for Architectural Surveyors and others engaged in Building:

Containing Formulæ useful in Designing Builder's work, Table of Weights, of the materials used in Building, Memoranda connected with Builders' work, Mensuration, the Practice of Builders' Measurement, Contracts of Labor, Valuation of Property, Summary of the Practice in Dilapidation, etc., etc. By J. F. HURST, C. E. Second edition, pocket-book form, full bound $2.00

JERVIS.—Railway Property:

A Treatise on the Construction and Management of Railways; designed to afford useful knowledge, in the popular style, to the holders of this class of property; as well as Railway Managers, Officers, and Agents. By JOHN B. JERVIS, late Chief Engineer of the Hudson River Railroad, Croton Aqueduct, etc. In one vol., 12mo., cloth $2.00

JOHNSTON.—Instructions for the Analysis of Soils, Limestones, and Manures.

By J. F. W. JOHNSTON. 12mo. . . . . . .

KEENE.—A Hand-Book of Practical Gauging:
For the Use of Beginners, to which is added, A Chapter on Distillation, describing the process in operation at the Custom House for ascertaining the strength of wines. By JAMES B. KEENE, of H. M. Customs. 8vo. $1.25

KELLEY.—Speeches, Addresses, and Letters on Industrial and Financial Questions.
By Hon. WILLIAM D. KELLEY, M. C. In one volume, 544 pages, 8vo. $3.00

KENTISH.—A Treatise on a Box of Instruments,
And the Slide Rule; with the Theory of Trigonometry and Logarithms, including Practical Geometry, Surveying, Measuring of Timber, Cask and Malt Gauging, Heights, and Distances. By THOMAS KENTISH. In one volume. 12mo. $1.25

KOBELL.—ERNI.—Mineralogy Simplified:
A short Method of Determining and Classifying Minerals, by means of simple Chemical Experiments in the Wet Way. Translated from the last German Edition of F. VON KOBELL, with an Introduction to Blow-pipe Analysis and other additions. By HENRI ERNI, M. D., late Chief Chemist, Department of Agriculture, author of "Coal Oil and Petroleum." In one volume, 12mo. $2.50

LANDRIN.—A Treatise on Steel:
Comprising its Theory, Metallurgy, Properties, Practical Working, and Use. By M. H. C. LANDRIN, Jr., Civil Engineer. Translated from the French, with Notes, by A. A. FESQUET, Chemist and Engineer. With an Appendix on the Bessemer and the Martin Processes for Manufacturing Steel, from the Report of Abram S. Hewitt, United States Commissioner to the Universal Exposition, Paris, 1867. In one volume, 12mo. $3.00

LARKIN.—The Practical Brass and Iron Founder's Guide:
A Concise Treatise on Brass Founding, Moulding, the Metals and their Alloys, etc.: to which are added Recent Improvements in the Manufacture of Iron, Steel by the Bessemer Process, etc., etc. By JAMES LARKIN, late Conductor of the Brass Foundry Department in Reany, Neafie & Co's. Penn Works, Philadelphia. Fifth edition, revised, with Extensive additions. In one volume, 12mo. . . $2.25

LEAVITT.—Facts about Peat as an Article of Fuel:
With Remarks upon its Origin and Composition, the Localities in which it is found, the Methods of Preparation and Manufacture, and the various Uses to which it is applicable; together with many other matters of Practical and Scientific Interest. To which is added a chapter on the Utilization of Coal Dust with Peat for the Production of an Excellent Fuel at Moderate Cost, specially adapted for Steam Service. By T. H. LEAVITT. Third edition. 12mo. . . . $1.75

LEROUX, C.—A Practical Treatise on the Manufacture of Worsteds and Carded Yarns:
Comprising Practical Mechanics, with Rules and Calculations applied to Spinning; Sorting, Cleaning, and Scouring Wools; the English and French methods of Combing, Drawing, and Spinning Worsteds and Manufacturing Carded Yarns. Translated from the French of CHARLES LEROUX, Mechanical Engineer, and Superintendent of a Spinning Mill, by HORATIO PAINE, M. D., and A. A. FESQUET, Chemist and Engineer. Illustrated by 12 large Plates. To which is added an Appendix, containing extracts from the Reports of the International Jury, and of the Artisans selected by the Committee appointed by the Council of the Society of Arts, London, on Woollen and Worsted Machinery and Fabrics, as exhibited in the Paris Universal Exposition, 1867. 8vo., cloth. $5.00

LESLIE (Miss).—Complete Cookery:
Directions for Cookery in its Various Branches. By MISS LESLIE. 60th thousand. Thoroughly revised, with the addition of New Receipts. In one volume, 12mo., cloth. $1.50

LESLIE (Miss).—Ladies' House Book:
A Manual of Domestic Economy. 20th revised edition. 12mo., cloth.

LESLIE (Miss).—Two Hundred Receipts in French Cookery.
Cloth, 12mo.

LIEBER.—Assayer's Guide:
Or, Practical Directions to Assayers, Miners, and Smelters, for the Tests and Assays, by Heat and by Wet Processes, for the Ores of all the principal Metals, of Gold and Silver Coins and Alloys, and of Coal, etc. By OSCAR M. LIEBER. 12mo., cloth. . . $1.25

LOTH.—The Practical Stair Builder:
A Complete Treatise on the Art of Building Stairs and Hand-Rails, Designed for Carpenters, Builders, and Stair-Builders. Illustrated with Thirty Original Plates. By C. EDWARD LOTH, Professional Stair-Builder. One large 4to. volume. $10.00

LOVE.—The Art of Dyeing, Cleaning, Scouring, and Finishing, on the Most Approved English and French Methods:
Being Practical Instructions in Dyeing Silks, Woollens, and Cottons, Feathers, Chips, Straw, etc. Scouring and Cleaning Bed and Window Curtains, Carpets, Rugs, etc. French and English Cleaning, any Color or Fabric of Silk, Satin, or Damask. By THOMAS LOVE, a Working Dyer and Scourer. Second American Edition, to which are added General Instructions for the Use of Aniline Colors. In one volume, 8vo., 343 pages. $5.00

MAIN and BROWN.—Questions on Subjects Connected with the Marine Steam-Engine:
And Examination Papers: with Hints for their Solution. By Thomas J. Main, Professor of Mathematics, Royal Naval College, and Thomas Brown, Chief Engineer, R. N. 12mo., cloth. . . . $1.50

MAIN and BROWN.—The Indicator and Dynamometer:
With their Practical Applications to the Steam-Engine. By Thomas J. Main, M. A. F. R., Assistant Professor Royal Naval College, Portsmouth, and Thomas Brown, Assoc. Inst. C. E., Chief Engineer, R. N., attached to the Royal Naval College. Illustrated. From the Fourth London Edition. 8vo. $1.50

MAIN and BROWN.—The Marine Steam-Engine.
By Thomas J. Main, F. R.; Assistant S. Mathematical Professor at the Royal Naval College, Portsmouth, and Thomas Brown, Assoc. Inst. C. E., Chief Engineer R. N. Attached to the Royal Naval College. Authors of "Questions connected with the Marine Steam-Engine," and the "Indicator and Dynamometer." With numerous Illustrations. In one volume, 8vo. $5.00

MARTIN.—Screw-Cutting Tables, for the Use of Mechanical Engineers:
Showing the Proper Arrangement of Wheels for Cutting the Threads of Screws of any required Pitch; with a Table for Making the Universal Gas-Pipe Thread and Taps. By W. A. Martin, Engineer. 8vo. 50

Mechanics' (Amateur) Workshop:
A treatise containing plain and concise directions for the manipulation of Wood and Metals, including Casting, Forging, Brazing, Soldering, and Carpentry. By the author of the "Lathe and its Uses." Third edition. Illustrated. 8vo. $3.00

MOLESWORTH.—Pocket-Book of Useful Formulæ and Memoranda for Civil and Mechanical Engineers.
By Guilford L. Molesworth, Member of the Institution of Civil Engineers, Chief Resident Engineer of the Ceylon Railway. Second American, from the Tenth London Edition. In one volume, full bound in pocket-book form. $2.00

NAPIER.—A System of Chemistry Applied to Dyeing.
By James Napier, F. C. S. A New and Thoroughly Revised Edition. Completely brought up to the present state of the Science, including the Chemistry of Coal Tar Colors, by A. A. Fesquet, Chemist and Engineer. With an Appendix on Dyeing and Calico Printing, as shown at the Universal Exposition, Paris, 1867. Illustrated. In one volume, 8vo., 422 pages. $5.00

NAPIER.—Manual of Electro-Metallurgy:
Including the Application of the Art to Manufacturing Processes. By JAMES NAPIER. Fourth American, from the Fourth London edition, revised and enlarged. Illustrated by engravings. In one vol., 8vo. $2.00

NASON.—Table of Reactions for Qualitative Chemical Analysis.
By HENRY B. NASON, Professor of Chemistry in the Rensselaer Polytechnic Institute, Troy, New York. Illustrated by Colors. . 63

NEWBERY.—Gleanings from Ornamental Art of every style:
Drawn from Examples in the British, South Kensington, Indian, Crystal Palace, and other Museums, the Exhibitions of 1851 and 1862, and the best English and Foreign works. In a series of one hundred exquisitely drawn Plates, containing many hundred examples. By ROBERT NEWBERY. 4to. $12.50

NICHOLSON.—A Manual of the Art of Bookbinding:
Containing full instructions in the different Branches of Forwarding, Gilding, and Finishing. Also, the Art of Marbling Book-edges and Paper. By JAMES B. NICHOLSON. Illustrated. 12mo., cloth. $2.25

NICHOLSON.—The Carpenter's New Guide:
A Complete Book of Lines for Carpenters and Joiners. By PETER NICHOLSON. The whole carefully and thoroughly revised by H. K. DAVIS, and containing numerous new and improved and original Designs for Roofs, Domes, etc. By SAMUEL SLOAN, Architect. Illustrated by 80 plates. 4to. . . . .

NORRIS.—A Hand-book for Locomotive Engineers and Machinists:
Comprising the Proportions and Calculations for Constructing Locomotives; Manner of Setting Valves; Tables of Squares, Cubes, Areas, etc., etc. By SEPTIMUS NORRIS, Civil and Mechanical Engineer. New edition. Illustrated. 12mo., cloth. $1.50

NYSTROM.—On Technological Education, and the Construction of Ships and Screw Propellers:
For Naval and Marine Engineers. By JOHN W. NYSTROM, late Acting Chief Engineer, U. S. N. Second edition, revised with additional matter. Illustrated by seven engravings. 12mo. . . $1.50

O'NEILL.—A Dictionary of Dyeing and Calico Printing:
Containing a brief account of all the Substances and Processes in use in the Art of Dyeing and Printing Textile Fabrics; with Practical Receipts and Scientific Information. By CHARLES O'NEILL, Analytical Chemist; Fellow of the Chemical Society of London; Member of the Literary and Philosophical Society of Manchester; Author of "Chemistry of Calico Printing and Dyeing." To which is added an Essay on Coal Tar Colors and their application to Dyeing and Calico Printing. By A. A. FESQUET, Chemist and Engineer. With an Appendix on Dyeing and Calico Printing, as shown at the Universal Exposition, Paris, 1867. In one volume, 8vo., 491 pages. . $5.00

ORTON.—Underground Treasures:
How and Where to Find Them. A Key for the Ready Determination of all the Useful Minerals within the United States. By JAMES ORTON, A. M. Illustrated, 12mo. $1.50

OSBORN.—American Mines and Mining:
Theoretically and Practically Considered. By Prof. H. S. OSBORN. Illustrated by numerous engravings. 8vo. (*In preparation.*)

OSBORN.—The Metallurgy of Iron and Steel:
Theoretical and Practical in all its Branches; with special reference to American Materials and Processes. By H. S. OSBORN, LL. D., Professor of Mining and Metallurgy in Lafayette College, Easton, Pennsylvania. Illustrated by numerous large folding plates and wood-engravings. 8vo. $15.00

OVERMAN.—The Manufacture of Steel:
Containing the Practice and Principles of Working and Making Steel. A Handbook for Blacksmiths and Workers in Steel and Iron, Wagon Makers, Die Sinkers, Cutlers, and Manufacturers of Files and Hardware, of Steel and Iron, and for Men of Science and Art. By FREDERICK OVERMAN, Mining Engineer, Author of the "Manufacture of Iron," etc. A new, enlarged, and revised Edition. By A. A. FESQUET, Chemist and Engineer. $1.50

OVERMAN.—The Moulder and Founder's Pocket Guide:
A Treatise on Moulding and Founding in Green-sand, Dry-sand, Loam, and Cement; the Moulding of Machine Frames, Mill-gear, Hollow-ware, Ornaments, Trinkets, Bells, and Statues; Description of Moulds for Iron, Bronze, Brass, and other Metals; Plaster of Paris, Sulphur, Wax, and other articles commonly used in Casting; the Construction of Melting Furnaces, the Melting and Founding of Metals; the Composition of Alloys and their Nature. With an Appendix containing Receipts for Alloys, Bronze, Varnishes and Colors for Castings; also, Tables on the Strength and other qualities of Cast Metals. By FREDERICK OVERMAN, Mining Engineer, Author of "The Manufacture of Iron." With 42 Illustrations. 12mo. $1.50

Painter, Gilder, and Varnisher's Companion:
Containing Rules and Regulations in everything relating to the Arts of Painting, Gilding, Varnishing, Glass-Staining, Graining, Marbling, Sign-Writing, Gilding on Glass, and Coach Painting and Varnishing; Tests for the Detection of Adulterations in Oils, Colors, etc.; and a Statement of the Diseases to which Painters are peculiarly liable, with the Simplest and Best Remedies. Sixteenth Edition. Revised, with an Appendix. Containing Colors and Coloring—Theoretical and Practical. Comprising descriptions of a great variety of Additional Pigments, their Qualities and Uses, to which are added, Dryers, and Modes and Operations of Painting, etc. Together with Chevreul's Principles of Harmony and Contrast of Colors. 12mo., cloth. $1.50

HENRY CAREY BAIRD'S CATALOGUE.   17

PALLETT.—The Miller's, Millwright's, and Engineer's Guide.
By HENRY PALLETT. Illustrated. In one volume, 12mo. • $3.00

PERCY.—The Manufacture of Russian Sheet-Iron.
By JOHN PERCY, M.D., F.R.S., Lecturer on Metallurgy at the Royal School of Mines, and to The Advanced Class of Artillery Officers at the Royal Artillery Institution, Woolwich; Author of "Metallurgy." With Illustrations. 8vo., paper. . . . . . . 50 cts.

PERKINS.—Gas and Ventilation.
Practical Treatise on Gas and Ventilation. With Special Relation to Illuminating, Heating, and Cooking by Gas. Including Scientific Helps to Engineer-students and others. With Illustrated Diagrams. By E. E. PERKINS. 12mo., cloth. $1.25

PERKINS and STOWE.—A New Guide to the Sheet-iron and Boiler Plate Roller:
Containing a Series of Tables showing the Weight of Slabs and Piles to produce Boiler Plates, and of the Weight of Piles and the Sizes of Bars to produce Sheet-iron; the Thickness of the Bar Gauge in decimals; the Weight per foot, and the Thickness on the Bar or Wire Gauge of the fractional parts of an inch; the Weight per sheet, and the Thickness on the Wire Gauge of Sheet-iron of various dimensions to weigh 112 lbs. per bundle; and the conversion of Short Weight into Long Weight, and Long Weight into Short. Estimated and collected by G. H. PERKINS and J. G. STOWE. $2.50

PHILLIPS and DARLINGTON.—Records of Mining and Metallurgy;
Or Facts and Memoranda for the use of the Mine Agent and Smelter. By J. ARTHUR PHILLIPS, Mining Engineer, Graduate of the Imperial School of Mines, France, etc., and JOHN DARLINGTON. Illustrated by numerous engravings. In one volume, 12mo. . . $1.50

PROTEAUX.—Practical Guide for the Manufacture of Paper and Boards.
By A. PROTEAUX, Civil Engineer, and Graduate of the School of Arts and Manufactures, and Director of Thiers' Paper Mill, Puy-de-Dôme. With additions, by L. S. LE NORMAND. Translated from the French, with Notes, by HORATIO PAINE, A. B., M. D. To which is added a Chapter on the Manufacture of Paper from Wood in the United States, by HENRY T. BROWN, of the "American Artisan." Illustrated by six plates, containing Drawings of Raw Materials, Machinery, Plans of Paper-Mills, etc., etc. 8vo. $10.00

REGNAULT.—Elements of Chemistry.
By M. V. REGNAULT. Translated from the French by T. FORREST BETTON, M. D., and edited, with Notes, by JAMES C. BOOTH, Melter and Refiner U. S. Mint, and WM. L. FABER, Metallurgist and Mining Engineer. Illustrated by nearly 700 wood engravings. Comprising nearly 1500 pages. In two volumes, 8vo., cloth. . . . $7.50

REID.—A Practical Treatise on the Manufacture of Portland Cement:
By HENRY REID, C. E. To which is added a Translation of M. A. Lipowitz's Work, describing a New Method adopted in Germany for Manufacturing that Cement, by W. F. REID. Illustrated by plates and wood engravings. 8vo. $5.00

RIFFAULT, VERGNAUD, and TOUSSAINT.—A Practical Treatise on the Manufacture of Varnishes.
By MM. RIFFAULT, VERGNAUD, and TOUSSAINT. Revised and Edited by M. F. MALEPEYRE and Dr. EMIL WINCKLER. Illustrated. In one volume, 8vo. (*In preparation.*)

RIFFAULT, VERGNAUD, and TOUSSAINT.—A Practical Treatise on the Manufacture of Colors for Painting:
Containing the best Formulæ and the Processes the Newest and in most General Use. By MM. RIFFAULT, VERGNAUD, and TOUSSAINT. Revised and Edited by M. F. MALEPEYRE and Dr. EMIL WINCKLER. Translated from the French by A. A. FESQUET, Chemist and Engineer. Illustrated by Engravings. In one volume, 650 pages, 8vo.
$7.50

ROBINSON.—Explosions of Steam Boilers:
How they are Caused, and how they may be Prevented. By J. R. ROBINSON, Steam Engineer. 12mo. $1.25

ROPER.—A Catechism of High Pressure or Non-Condensing Steam-Engines:
Including the Modelling, Constructing, Running, and Management of Steam Engines and Steam Boilers. With Illustrations. By STEPHEN ROPER, Engineer. Full bound tucks . . . $2.00

ROSELEUR.—Galvanoplastic Manipulations:
A Practical Guide for the Gold and Silver Electro-plater and the Galvanoplastic Operator. Translated from the French of ALFRED ROSELEUR, Chemist, Professor of the Galvanoplastic Art, Manufacturer of Chemicals, Gold and Silver Electro-plater. By A. A. FESQUET, Chemist and Engineer. Illustrated by over 127 Engravings on wood. 8vo., 495 pages. $6.00

☞ *This Treatise is the fullest and by far the best on this subject ever published in the United States.*

SCHINZ.—Researches on the Action of the Blast Furnace.
By CHARLES SCHINZ. Translated from the German with the special permission of the Author by WILLIAM H. MAW and MORITZ MULLER. With an Appendix written by the Author expressly for this edition. Illustrated by seven plates, containing 28 figures. In one volume, 12mo. $4.00

SHAW.—Civil Architecture:
Being a Complete Theoretical and Practical System of Building, containing the Fundamental Principles of the Art. By EDWARD SHAW, Architect. To which is added a Treatise on Gothic Architecture, etc. By THOMAS W. SILLOWAY and GEORGE M. HARDING, Architects. The whole illustrated by One Hundred and Two quarto plates finely engraved on copper. Eleventh Edition. 4to., cloth. . $10.00

SHUNK.—A Practical Treatise on Railway Curves and Location, for Young Engineers.
By WILLIAM F. SHUNK, Civil Engineer. 12mo. . . $2.00

SLOAN.—American Houses:
A variety of Original Designs for Rural Buildings. Illustrated by 26 colored Engravings, with Descriptive References. By SAMUEL SLOAN, Architect, author of the "Model Architect," etc., etc. 8vo. $1.50

SMEATON.—Builder's Pocket Companion:
Containing the Elements of Building, Surveying, and Architecture; with Practical Rules and Instructions connected with the subject. By A. C. SMEATON, Civil Engineer, etc. In one volume, 12mo. $1.50

SMITH.—A Manual of Political Economy.
By E. PESHINE SMITH. A new Edition, to which is added a full Index. 12mo., cloth. $1.25

SMITH.—Parks and Pleasure Grounds:
Or Practical Notes on Country Residences, Villas, Public Parks, and Gardens. By CHARLES H. J. SMITH, Landscape Gardener and Garden Architect, etc., etc. 12mo. $2.25

SMITH.—The Dyer's Instructor:
Comprising Practical Instructions in the Art of Dyeing Silk, Cotton, Wool, and Worsted, and Woollen Goods: containing nearly 800 Receipts. To which is added a Treatise on the Art of Padding; and the Printing of Silk Warps, Skeins, and Handkerchiefs, and the various Mordants and Colors for the different styles of such work. By DAVID SMITH, Pattern Dyer. 12mo., cloth. . . . $3.00

SMITH.—The Dyer's Instructor:
Comprising Practical Instructions in the Art of Dyeing Silk, Cotton, Wool, and Worsted and Woollen Goods. Third Edition, with many additional Receipts for Dyeing the New Alkaline Blues and Night Greens, *with Dyed Patterns affixed*. 12mo., pp. 394, cloth. . $10.50

STEWART.—The American System.
Speeches on the Tariff Question, and on Internal Improvements, principally delivered in the House of Representatives of the United States. By ANDREW STEWART, late M. C. from Pennsylvania. With a Portrait, and a Biographical Sketch. In one volume, 8vo., 407 pages. $3.00

STOKES.—Cabinet-maker's and Upholsterer's Companion:
Comprising the Rudiments and Principles of Cabinet-making and Upholstery, with Familiar Instructions, illustrated by Examples for attaining a Proficiency in the Art of Drawing, as applicable to Cabinet-work; the Processes of Veneering, Inlaying, and Buhl-work; the Art of Dyeing and Staining Wood, Bone, Tortoise Shell, etc. Directions for Lackering, Japanning, and Varnishing; to make French Polish; to prepare the Best Glues, Cements, and Compositions, and a number of Receipts particularly useful for workmen generally. By J. STOKES. In one volume, 12mo. With Illustrations. . $1.25

Strength and other Properties of Metals:
Reports of Experiments on the Strength and other Properties of Metals for Cannon. With a Description of the Machines for testing Metals, and of the Classification of Cannon in service. By Officers of the Ordnance Department U. S. Army. By authority of the Secretary of War. Illustrated by 25 large steel plates. In one volume, 4to. . $10.00

SULLIVAN.—Protection to Native Industry.
By Sir EDWARD SULLIVAN, Baronet, author of "Ten Chapters on Social Reforms." In one volume, 8vo. $1.50

Tables Showing the Weight of Round, Square, and Flat Bar Iron, Steel, etc.,
By Measurement. Cloth. 63

TAYLOR.—Statistics of Coal:
Including Mineral Bituminous Substances employed in Arts and Manufactures; with their Geographical, Geological, and Commercial Distribution and Amount of Production and Consumption on the American Continent. With Incidental Statistics of the Iron Manufacture. By R. C. TAYLOR. Second edition, revised by S. S. HALDEMAN. Illustrated by five Maps and many wood engravings. 8vo., cloth. $10.00

TEMPLETON.—The Practical Examinator on Steam and the Steam-Engine:
With Instructive References relative thereto, arranged for the Use of Engineers, Students, and others. By WM. TEMPLETON, Engineer. 12mo. $1.25

THOMAS.—The Modern Practice of Photography.
By R. W. THOMAS, F. C. S. 8vo., cloth. 75

THOMSON.—Freight Charges Calculator.
By ANDREW THOMSON, Freight Agent. 24mo. . . . $1.25

TURNING: Specimens of Fancy Turning Executed on the Hand or Foot Lathe:
With Geometric, Oval, and Eccentric Chucks, and Elliptical Cutting Frame. By an Amateur. Illustrated by 30 exquisite Photographs. 4to. $3.00

Turner's (The) Companion:
Containing Instructions in Concentric, Elliptic, and Eccentric Turning: also various Plates of Chucks, Tools, and Instruments; and Directions for using the Eccentric Cutter, Drill, Vertical Cutter, and Circular Rest; with Patterns and Instructions for working them. A new edition in one volume, 12mo. $1.50

URBIN.—BRULL.—A Practical Guide for Puddling Iron and Steel.
By ED. URBIN, Engineer of Arts and Manufactures. A Prize Essay read before the Association of Engineers, Graduate of the School of Mines, of Liege, Belgium, at the Meeting of 1865–6. To which is added A COMPARISON OF THE RESISTING PROPERTIES OF IRON AND STEEL. By A. BRULL. Translated from the French by A. A. FESQUET, Chemist and Engineer. In one volume, 8vo. $1.00

VAILE.—Galvanized Iron Cornice-Worker's Manual:
Containing Instructions in Laying out the Different Mitres, and Making Patterns for all kinds of Plain and Circular Work. Also, Tables of Weights, Areas and Circumferences of Circles, and other Matter calculated to Benefit the Trade. By CHARLES A. VAILE, Superintendent "Richmond Cornice Works," Richmond, Indiana. Illustrated by 21 Plates. In one volume, 4to. $5.00

VILLE.—The School of Chemical Manures:
Or, Elementary Principles in the Use of Fertilizing Agents. From the French of M. GEORGE VILLE, by A. A. FESQUET, Chemist and Engineer. With Illustrations. In one volume, 12 mo.   .   . $1.25

VOGDES.—The Architect's and Builder's Pocket Companion and Price Book:
Consisting of a Short but Comprehensive Epitome of Decimals, Duodecimals, Geometry and Mensuration; with Tables of U. S. Measures, Sizes, Weights, Strengths, etc., of Iron, Wood, Stone, and various other Materials, Quantities of Materials in Given Sizes, and Dimensions of Wood, Brick, and Stone; and a full and complete Bill of Prices for Carpenter's Work; also, Rules for Computing and Valuing Brick and Brick Work, Stone Work, Painting, Plastering, etc. By FRANK W. VOGDES, Architect. Illustrated. Full bound in pocket-book form. $2.00
Bound in cloth. 1.50

WARN.—The Sheet-Metal Worker's Instructor:
For Zinc, Sheet-Iron, Copper, and Tin-Plate Workers, etc. Containing a selection of Geometrical Problems; also, Practical and Simple Rules for describing the various Patterns required in the different branches of the above Trades. By REUBEN H. WARN, Practical Tin-plate Worker. To which is added an Appendix, containing Instructions for Boiler Making, Mensuration of Surfaces and Solids, Rules for Calculating the Weights of different Figures of Iron and Steel, Tables of the Weights of Iron, Steel, etc. Illustrated by 32 Plates and 37 Wood Engravings. 8vo. $3.00

WATSON.—A Manual of the Hand-Lathe:
Comprising Concise Directions for working Metals of all kinds, Ivory, Bone and Precious Woods; Dyeing, Coloring, and French Polishing; Inlaying by Veneers, and various methods practised to produce Elaborate work with Dispatch, and at Small Expense. By EGBERT P. WATSON, late of "The Scientific American," Author of "The Modern Practice of American Machinists and Engineers." Illustrated by 78 Engravings. $1.50

WATSON.—The Modern Practice of American Machinists and Engineers:
Including the Construction, Application, and Use of Drills, Lathe Tools, Cutters for Boring Cylinders, and Hollow Work Generally, with the most Economical Speed for the same; the Results verified by Actual Practice at the Lathe, the Vice, and on the Floor. Together with Workshop Management, Economy of Manufacture, the Steam-Engine, Boilers, Gears, Belting, etc., etc. By EGBERT P. WATSON, late of the "Scientific American." Illustrated by 86 Engravings. In one volume, 12mo. $2.50

WATSON.—The Theory and Practice of the Art of Weaving by Hand and Power:
With Calculations and Tables for the use of those connected with the Trade. By JOHN WATSON, Manufacturer and Practical Machine Maker. Illustrated by large Drawings of the best Power Looms. 8vo. $7.50

WEATHERLY.—Treatise on the Art of Boiling Sugar, Crystallizing, Lozenge-making, Comfits, Gum Goods.
12mo. $2.00

WILL.—Tables for Qualitative Chemical Analysis.
By Professor HEINRICH WILL, of Giessen, Germany. Seventh edition. Translated by CHARLES F. HIMES, Ph. D., Professor of Natural Science, Dickinson College, Carlisle, Pa. . . . $1.50

WILLIAMS.—On Heat and Steam:
Embracing New Views of Vaporization, Condensation, and Explosions. By CHARLES WYE WILLIAMS, A. I. C. E. Illustrated. 8vo. $3.50

WOHLER.—A Hand-Book of Mineral Analysis.
By F. WOHLER, Professor of Chemistry in the University of Göttingen. Edited by HENRY B. NASON, Professor of Chemistry in the Rensselaer Polytechnic Institute, Troy, New York. Illustrated. In one volume, 12mo. $3 00

WORSSAM.—On Mechanical Saws:
From the Transactions of the Society of Engineers, 1869. By S. W. WORSSAM, Jr. Illustrated by 18 large plates. 8vo. . . $5.00

RECENT ADDITIONS TO OUR LIST.

AUERBACH.—Anthracen: Its Constitution, Properties, Manufacture, and Derivatives, including Artificial Alizarin, Anthrapurpurin, with their applications in Dyeing and Printing. By G. AUERBACH. Translated and edited by WM. CROOKES, F. R. S. 8vo. $5.00

BECKETT.—Treatise on Clocks, Watches and Bells. By SIR EDMUND BECKETT, Bart. Illustrated. 12mo. . $1.75

BARLOW.—The History and Principles of Weaving, by Hand and by Power. Several Hundred Illustrations. 8vo. $10.00

BOURNE.—Recent Improvements in the Steam Engine. By JOHN BOURNE, C. E. Illustrated. 16mo. . . . $1.50

CLARK.—Fuel: Its Combustion and Economy. By D. KINNEAR CLARK, C. E. 144 Engravings. 12mo. . $2.25

CRISTIANI.—Perfumery and Kindred Arts. By R. S. CRISTIANI. 8vo. $5.00

COLLENS.—The Eden of Labor, or the Christian Utopia. 12mo. Paper, $1.00; Cloth, ... $1.25

CUPPER.—The Universal Stair Builder. Illustrated by 29 plates. 4to. $5.00

COOLEY.—A Complete Practical Treatise on Perfumery. By A. J. COOLEY. 12mo. $1.50

DAVIDSON.—A Practical Manual of House Painting, Graining, Marbling and Sign Writing: With 9 Colored Illustrations of Woods and Marbles, and many Wood Engravings. 12mo. $3.00

EDWARDS.—A Catechism of the Marine Steam Engine. By EMORY EDWARDS. Illustrated. 12mo. . . . $2.00

HASERICK.—The Secrets of he Art of Dyeing Wool, Cotton, and Linen: Including Bleaching and Coloring Wool and Cotton Hosiery and Random Yarns. By E. C. HASERICK. Illustrated by 323 Dyed Patterns of the Yarns or Fabrics. 8vo. $25.00

HENRY.—The Early and Later History of Petroleum. By J. T. HENRY. Illustrated. 8vo. $4.50

KELLOGG.—A New Monetary System.
By Ed. Kellogg. Fifth Edition. Edited by Mary Kellogg Putnam. 12mo. Paper, $1.00; Cloth, ... $1.50

KEMLO.—Watch Repairer's Hand-Book.
Illustrated. 12mo. $1.25

MORRIS.—Easy Rules for the Measurement of Earthworks by means of the Prismoidal Formula.
By Elwood Morris, C. E. 8vo. $1.50

McCULLOCH.—Distillation, Brewing and Malting.
By J. C. McCulloch. 12mo. $1.00

NEVILLE.—Hydraulic Tables, Co-Efficients, and Formulæ for Finding the Discharge of Water from Orifices, Notches, Weirs, Pipes, and Rivers.
Illustrated. 12mo. $5.00

NICOLLS.—The Railway Builder.
A Hand-book for Estimating the Probable Cost of American Railway Construction and Equipment. By Wm. J. Nicolls, C. E. Pocket-book Form. $2.00

NORMANDY.—The Commercial Hand-book of Chemical Analysis.
By H. M. Noad, Ph. D. 12mo. $5.00

PROCTOR.—A Pocket-Book of Useful Tables and Formulæ for Marine Engineers.
By Frank Proctor. Pocket-book Form. . . . $2.00

ROSE.—The Complete Practical Machinist:
Embracing Lathe Work, Vise Work, Drills and Drilling, Taps and Dies, Hardening and Tempering, the Making and Use of Tools, etc., etc. By Joshua Rose. 130 Illustrations. 12mo. . . $2.50

SLOAN.—Homestead Architecture.
By Samuel Sloan, Architect. 200 Engravings. 8vo. . $3.50

SYME.—Outlines of an Industrial Science.
By David Syme. 12mo. $2.00

WARE.—The Coachmaker's Illustrated Hand-Book.
Fully Illustrated. 8vo. $3.00

WIGHTWICK.—Hints to Young Architects.
Numerous Wood Cuts. 12mo. $2.00

WILSON.—First Principles of Political Economy.
12mo. $1.50

WILSON.—A Treatise on Steam Boilers, their Strength, Construction, and Economical Working.
By Robt. Wilson. Illustrated. 12mo. . . . $2.50

www.ingramcontent.com/pod-product-compliance
Lightning Source LLC
Chambersburg PA
CBHW020113170426
43199CB00009B/518